T0340108

NATURE-INSPIRED OPTIMIZATION ALGORITHMS FOR FUZZY CONTROLLED SERVO SYSTEMS

NATURE-INSPIRED OPTIMIZATION ALGORITHMS FOR FUZZY CONTROLLED SERVO SYSTEMS

RADU-EMIL PRECUP

RADU-CODRUT DAVID

Butterworth-Heinemann
An imprint of Elsevier

Butterworth-Heinemann is an imprint of Elsevier
The Boulevard, Langford Lane, Kidlington, Oxford OX5 1GB, United Kingdom
50 Hampshire Street, 5th Floor, Cambridge, MA 02139, United States

Library of Congress Cataloging-in-Publication Data
A catalog record for this book is available from the Library of Congress

British Library Cataloguing-in-Publication Data
A catalogue record for this book is available from the British Library
ISBN: 978-0-12-816358-0

For information on all Butterworth-Heinemann publications visit
our website at https://www.elsevier.com/books-and-journals

Working together
to grow libraries in
developing countries

www.elsevier.com • www.bookaid.org

Publisher: Mara Conner
Acquisition Editor: Sonnini R. Yura
Editorial Project Manager: Gabriela D. Capille
Production Project Manager: Kamesh Ramajogi
Cover Designer: Mark Rogers

Typeset by SPi Global, India

Contents

1. **Introduction**   **1**
   1.1 Motivation and overview   1
   1.2 Fuzzy sets, operations on fuzzy sets, fuzzy relations   7
   1.3 Information processing in fuzzy controllers   13
   1.4 Fuzzy controllers and design approaches   31
   1.5 Control system models and definitions of optimization problems   41
   References   49

2. **Nature-inspired algorithms for the optimal tuning of fuzzy controllers**   **55**
   2.1 Particle swarm optimization algorithms   55
   2.2 Gravitational search algorithms   62
   2.3 Charged system search algorithms   67
   2.4 Gray wolf optimizer algorithms   72
   References   78

3. **Adaptive nature-inspired algorithms for the optimal tuning of fuzzy controllers**   **81**
   3.1 Adaptive gravitational search algorithms   81
   3.2 Adaptive charged system search algorithms   86
   3.3 Fuzzy logic-based adaptive gravitational search algorithms   92
   References   99

4. **Hybrid nature-inspired algorithms for the optimal tuning of fuzzy controllers**   **103**
   4.1 Hybrid particle swarm optimization-gravitational search algorithms   103
   4.2 Hybrid gray wolf optimization-particle swarm optimization algorithms   108
   References   113

5. **Conclusions**   **115**
   5.1 Performance comparison of nature-inspired optimization algorithms in the optimal tuning of Takagi-Sugeno proportional-integral fuzzy controllers   115
   5.2 A sample of experimental results   119
   5.3 Extensions to proportional-integral-derivative fuzzy controllers   121
   5.4 Extensions to type-2 fuzzy controllers   123

5.5 Extensions to tensor product-based model transformation controllers   124

5.6 Extensions to evolving fuzzy systems and controllers   127

5.7 Perspectives of nature-inspired algorithms applied to fuzzy control   131

References   134

Index   *139*

CHAPTER 1

Introduction

Contents

1.1 Motivation and overview	1
1.2 Fuzzy sets, operations on fuzzy sets, fuzzy relations	7
1.3 Information processing in fuzzy controllers	13
1.3.1 The fuzzification module	13
1.3.2 The inference module	17
1.3.3 The defuzzification module	24
1.3.4 Takagi-Sugeno fuzzy models and Tsukamoto fuzzy models	28
1.4 Fuzzy controllers and design approaches	31
1.4.1 Fuzzy controllers without dynamics	31
1.4.2 Fuzzy controllers with dynamics	33
1.5 Control system models and definitions of optimization problems	41
References	49

Abstract

This chapter starts with the motivation and overview of this book. Information on fuzzy sets, operations on fuzzy sets and fuzzy relations are offered next to be understood and used in the fuzzy controller design and tuning. The fuzzy controller structure and design approaches are formulated for Mamdani and Takagi-Sugeno fuzzy controllers. The main subsystems of the fuzzy control systems for servo systems, namely the process and the proportional-integral fuzzy controller, are described by means structures and models. Four optimization problems that target the optimal tuning of fuzzy controllers for servo systems are defined.

Keywords: Fuzzy sets, Fuzzy relations, Fuzzy controllers, Linguistic terms, Linguistic variables, Position control, Servo systems

1.1 Motivation and overview

The classical engineering approaches to characterize real–world problems are essentially qualitative and quantitative ones, based on more or less accurate mathematical modeling. In such approaches expressions such as "medium temperature," "large humidity," "small pressure," "very large speed," related to the variables specific to the behavior of a **controlled process**

(**CP**), are subjected to relatively difficult quantitative interpretations. That happens because "classical" automation handles variables/information processed with well-specified numerical values. In this context the elaboration of the control strategy and its implementation in the control equipment require an as accurate as possible quantitative modeling of the CP. Advanced control strategies (adaptive, predictive, learning, or variable structure ones) require even the permanent reassessment of the models and of the values of the parameters characterizing these (parametric) models.

Process control based on **fuzzy set (FS) theory** or on **fuzzy logic**—referred to as **fuzzy control** or **fuzzy logic control**—is more pragmatic from this point of view. The reason for that concerns the capability to take over and use a linguistic characterization of the quality of CP dynamics and to adapt this characterization as function of the conditions of CP operation.

L.A. Zadeh set the basics of FS theory by a paper (Zadeh, 1965) that firstly seemed to be only mathematical entertainment. The boom in the 1970s in computer science opened the first prospects for practical applications of the meanwhile built theory in the field of process control/automatic control and these first applications belong to **E.H. Mamdani** and coworkers (Mamdani, 1974; Mamdani and Assilian, 1975). The reference application of fuzzy control deals with the use of some "special" controllers based on FS theory, **fuzzy controllers**, for cement kiln control (Holmblad and Ostergaard, 1982). In the 1980s in Japan, United States, and later in the Europe, the so-called *fuzzy boom* took place in the field of fuzzy control applications involving several domains that range from electrical household industry up to the control of vehicles, transportation systems, and robots. This is caused partly by the spectacular development of electronic technology and computer systems that enabled: (i) the manufacturing of circuits with very high speed of information processing, dedicated (by construction and usage) to a certain purpose including fuzzy information processing, and (ii) the development of computer-aided design programs, which allow the control system designer to use efficiently a large amount of information concerning the CP and the control equipment.

The applications of fuzzy control reported until now point out two *important aspects* related to this control strategy:
- In some situations (e.g., the control of process functional nonlinearities subject to difficult mathematical modeling or even the control of ill-defined processes), fuzzy control can be a **viable alternative** to classical, crisp control (conventional control).

- Compared to conventional control, fuzzy control can be strongly based and focused on the experience of a human operator, and a fuzzy controller can model more accurately this experience (in linguistic manner) vs a conventional controller.

The main features of fuzzy control are as follows:

- Fuzzy control employs the so-called **fuzzy controllers** (**FCs**) or **fuzzy logic controllers** (**FLCs**) ensuring a nonlinear input-output static map that can be influenced/modified based on designer's option.
- Fuzzy control can process several variables from the CP, hence it can be considered as belonging to the class of multiple input-multiple output (MIMO) systems with interactions, therefore the FC can be viewed as a **multiple input** controller (eventually, a multiple output one, too), similar to state-feedback controllers.
- FCs are controllers without dynamics, but the applications and performance of FCs and **fuzzy control systems** (**FCSs**) can be enlarged significantly by inserting dynamics (derivative and/or integral components) to fuzzy controller structures resulting in the so-called **fuzzy controllers with dynamics**.
- FCs are flexible with regard to the modification of the transfer features (by input-output static maps), thus ensuring the possibility to design a large variety of adaptive control system structures.

The control approach based on human experience is transposed in FCs by expressing the control requirements and elaborating the control signal (the control action) in terms of the natural IF-THEN rules which belong to the set of rules.

$$\ldots$$
$$\text{IF (antecedent) THEN (consequent)} \qquad (1.1)$$
$$\ldots$$

where the **antecedent** (**premise**) refers to the found-out situation concerning the CP evolution (usually compared with the desired one), and the **consequent** (**conclusion**) refers to the measures which should be taken–under the form of the control signal u–in order to achieve the desired dynamics. The set of rules (1.1) represents the **rule base** of the FC.

Some research results on the behavior of the human expert emphasize that the expert has a specific strongly nonlinear behavior accompanied by anticipative, derivative, integral, and predictive effects and by adaptation to the concrete operating conditions. Coloring the linguistic characterization of CP evolution (and, accordingly, of fuzzy mathematical

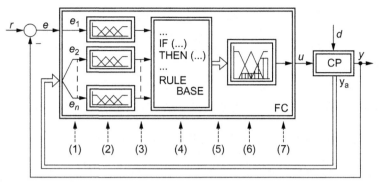

Fig. 1.1 Basic fuzzy control system structure. *(From Precup, R.-E., Hellendoorn, H., 2011. A survey on industrial applications of fuzzy control. Comput. Ind. 62(3), 213–226.)*

characterization) based on experience and translating it to the control signal elaboration and to the analysis of CP evolution (dynamics) will be characterized by parameters that enable the modification of FC features. From this point of view FCS can be regarded in the general framework of **intelligent control systems**.

The block diagram of principle (considered as classical in the literature) of an FCS is presented in Fig. 1.1 (Precup and Hellendoorn, 2011). The FCS is considered as a single input system with respect to the **reference input** (**setpoint**) r and as a single output system with respect to the **controlled output** y. The second input fed to the CP/FCS is the disturbance input d.

Fig. 1.1 also highlights the operating principle of an FC in its classical version that characterizes *Mamdani fuzzy controllers*, with the following **variables** and **modules**: (1) the crisp inputs, (2) the fuzzification module, (3) the fuzzified inputs, (4) the inference module, (5) the fuzzy conclusions, (6) the defuzzification module, and (7) the crisp output.

One essential particular feature of FCSs concerns the multiple interactions considered from the process to the controller by the auxiliary variables $\mathbf{y}_a$, gathered in the input vector $\mathbf{e}'$.

$$\mathbf{e}' = \begin{bmatrix} e\ \mathbf{y}_a^T \end{bmatrix} = \begin{bmatrix} e_1 & e_2 & \cdots & e_n \end{bmatrix}^T \tag{1.2}$$

These variables are direct or indirect inputs to the fuzzy controller and are also called **scheduling variables**. No matter how many inputs to the FC are, the FC should possess at least one input variable or one scheduling variable, e_1, which corresponds to the **control error** e

$$e_1 = e = r - y. \tag{1.3}$$

According to Fig. 1.1, the operating principle of a Mamdani FC involves the **sequence of operations** (a), (b), and (c):

(a)   The crisp input information–the measured variables, the reference input (the set point), the control error–is converted into fuzzy representation. This operation is called **fuzzification** of crisp information.

(b)   The fuzzified information is processed using the **rule base**, composed of the fuzzy IF-THEN rules referred to as fuzzy control rules of type (1.1) that must be well defined in order to control the given process. The principles to evaluate and process the rule base represent the **inference mechanism** or the **inference engine** and the result is the "fuzzy" form of the control signal u, namely the **fuzzy control signal**.

(c)   The fuzzy control signal must be converted into a crisp formulation, with well-specified physical nature, directly understandable and usable by the actuator in order to be capable of controlling the process. This operation is called **defuzzification**.

The three operations described briefly here characterize the three modules in the structure of an FC (Fig. 1.1), the fuzzification module (2), the inference module (4), and the defuzzification module (6). All three modules are supported by adequate databases.

A consistent way to achieve the performance specifications of FCS involves tuning the parameters of fuzzy controllers or fuzzy models with the aid of defined optimization problems with variables matching those parameters. The performance specifications are met by solving these optimization problems that ensure the optimal tuning of fuzzy controllers and fuzzy models. This process may lead to multi-objective optimization problems due to the complexity of the process and controller's structures and nonlinearities as the objective functions associated with the optimization problems could be non-convex and non-differentiable.

For control systems, the performance indices are usually expressed as empirical control system performance indices (e.g., overshoot, settling time, phase margin for linear systems, etc.). A common practice of achieving the desired performance specifications of FCS is to define them through **optimization problems** based on **objective functions**, which use **as variables the tuning parameters of the controller**, with appropriate constraints imposed. Optimal tuning parameters are reached by solving these optimization problems, which, in most cases, implies the minimization of the objective functions. The optimal tuning of fuzzy controllers is applied in this book in the context of the abovementioned systematic way to design and tune these nonlinear controllers.

Solving the optimization problems specific to the optimal tuning of fuzzy controllers is a complex task due to the complicated expressions of the objective functions and the risk of getting trapped in local minima situations. This depends on the processes and on the fuzzy controller structures, with suggestive examples discussed by Preitl and Precup (1997, 1999), Baranyi et al. (2003), Haber et al. (2003), Škrjanc and Blažič (2005), Castillo et al. (2007), Guechi et al. (2010), Linda and Manic (2011), Oh et al. (2011), Angelov and Yager (2012), Haidegger et al. (2012), Teodorescu (2012), Vaščák (2012), Precup et al. (2013c), Blažič et al. (2014), Mohammadzadeh et al. (2014), Pelusi et al. (2016), Johanyák (2017), Vrkalovic et al. (2017), and Gil et al. (2018). Useful surveys on fuzzy control are given by Kaynak et al. (2001), Sala et al. (2005), Feng (2006), Precup and Hellendoorn (2011), Castillo and Melin (2012, 2014), Guerra et al. (2015), Precup et al. (2015), and Qiu et al. (2016). In order to reduce the computational cost for minimizing these objective functions, *nature-inspired algorithms*, referred to also as *metaheuristic algorithms* or simply *metaheuristics*, can be used due to their derivative-free characteristic. Additionally, the use of these nature-inspired algorithms to optimally tune the parameters of fuzzy controllers can offer the following advantages: reduced running costs, transparency in the design, low-cost design and implementation, and gradient information replaced by actual objective function value.

The first objective of this book is to present in a unified structure from the point of view of a control engineer the essential aspects regarding fuzzy control in servo systems. **The second objective** is the optimal tuning of fuzzy controllers using nature-inspired optimization algorithms.

In the majority of applications an FC is used for direct feedback control or on the low level in hierarchical control system structures. However, it can be used on the supervisory level, for example, in adaptive control system structures. Nowadays fuzzy control is no longer only used to directly express the knowledge on the CP or, in other words, to do model-free fuzzy control. An FC can be calculated from a fuzzy model obtained in terms of system identification techniques, and thus it can be regarded in the framework of model-based fuzzy control. Most often used are

- Mamdani fuzzy controllers, referred to also as linguistic FCs, with either fuzzy consequents that represent type-I fuzzy systems according to the classifications given by Sugeno (1999) or singleton consequents that belong to type-II fuzzy systems. These FCs are usually used as direct closed-loop controllers, and transformation between fuzzy models are described by Kóczy (1996).

- Takagi-Sugeno (T-S) fuzzy controllers, also known as type-III fuzzy systems especially when affine consequents are employed, and are also used as supervisory controllers.
- The latest solutions for solving optimization problems comprising parameter tuning of fuzzy controllers and fuzzy models are built on nature-inspired optimization algorithms that include genetic algorithms (Onieva et al., 2012; Pelusi et al., 2016), simulated annealing (SA) (Precup et al., 2011; Precup et al., 2012), particle swarm optimization (PSO) (Oh et al., 2011; Precup et al., 2013b), gravitational search algorithms (GSAs) (David et al., 2013; Precup et al., 2013a; Mahmoodabadi and Danesh, 2017), charged system search (CSS) algorithms (Precup et al., 2014), ant colony optimization (Chang et al. 2012), gray wolf optimizer (GWO) (Noshadi et al., 2016; Precup et al., 2016, 2017a, b), and artificial bee colony (ABC) (Bouallègue et al., 2015). These solutions comprise both objectives of this book.

1.2 Fuzzy sets, operations on fuzzy sets, fuzzy relations

The essence of the fuzzy representation of information–often, but incorrectly called fuzzy information–is based on the introduction of a measure to characterize the membership of an element to a set. The **membership function (m.f.)** is used with this respect.

Definition 1.1 Let X be a **basic set (basic domain, universe, universe of discourse)** having the elements $x \in X$. The function μ_F, defined as

$$\mu_F : X \rightarrow [0, 1], \tag{1.4}$$

is called **m.f.** of the **FS** F, by which for each element $x \in X$ a value $\mu_F(x) \in [0, 1]$ is mapped, which characterizes the **membership degree** of x to X. A **FS** F (defined) on X, is completely defined by the following set of pairs:

$$F = \{(x, \mu_F(x)) \mid x \in X\}. \tag{1.5}$$

The universe of discourse can be a countable or discrete set. The following notation for the FS F is used in this situation:

$$F = \sum_{x \in X} \mu_F(x)/x, \tag{1.6}$$

where the symbol $\mu_F(x)/x = (x, \mu_F(x))$ represents the pair of discrete values "membership degree"/"crisp value" that belong to the universe, and is named **singleton**. The symbol $\sum$ characterizes the union of all pairs

$(x, \mu_F(x))$. The symbol $\sum$ is usually replaced by the symbol $\int$ if the universes are not countable or are continuous:

$$F = \int_X \mu_F(x)/x. \tag{1.7}$$

The representation of an FS by means of its m.f. can be carried out in terms of several approaches: (i) parametric using an analytic function that corresponds to the m.f., (ii) direct graphical by means of the graphics of the m.f., and (iii) discrete in terms of singletons for FSs with countable or discrete FSs. The approach (i) will be used in the sequel, and the following m.f.s are widely used in fuzzy control, with details on the other two approaches given by Preitl and Precup (1997, 1999):

- Trapezoidal m.f.:

$$\mu_F(x) = \begin{cases} 0, & \text{if } x < \alpha, \\ (x-\alpha)/(\beta-\alpha), & \text{if } x \in [\alpha, \beta], \\ 1, & \text{if } x \in [\beta, \gamma], \quad x \in R, \quad \alpha < \beta \leq \gamma < \delta, \\ (\delta-x)/(\delta-\gamma), & \text{if } x \in [\gamma, \delta], \\ 0, & \text{if } x > \delta, \end{cases} \tag{1.8}$$

- Triangular m.f., using $\beta = \gamma$ in Eq. (1.8),
- Singleton m.f.:

$$\mu_F(x) = \begin{cases} 1, & \text{if } x = \alpha, \\ 0, & \text{otherwise,} \end{cases} \quad x \in R, \tag{1.9}$$

- Gaussian m.f.:

$$\mu_F(x) = e^{-(x-\bar{x})^2/2\sigma^2}, \quad x \in R, \tag{1.10}$$

with the parameters $\bar{x}$—the center and $\sigma \neq 0$—the width,
- Generalized bell-shaped m.f.:

$$\mu_F(x) = \frac{1}{1 + |(x-\bar{x})/a|^{2b}}, \quad x \in R, \tag{1.11}$$

where $\bar{x}$ is the center, and the parameters $b \in R$, $a \neq 0$ set the bell width.

The parameter α associated to the singleton in Eq. (1.9) is also called **modal value** and treated by Galichet and Foulloy (1995). A singleton can be viewed as one representation of a crisp value that is equal to the model value.

The following **descriptors** are associated to the analytical characterization of an FS, defined as follows for the FS $F = \{(x, \mu_F(x)) \mid x \in X\}$:

- **the support of an FS**, Supp(F):

$$\text{Supp}(F) = \{x \mid x \in X, \ \mu_F(x) > 0\} \subset X, \tag{1.12}$$

- **the core of an FS**, $K(F)$:

$$K(F) = \{x \mid x \in X, \ \mu_F(x) = 1\} \subset X, \tag{1.13}$$

- **the height of an FS**, hgt(F):

$$\text{hgt}(F) = \max\{x \mid x \in X\} \in [0, 1]. \tag{1.14}$$

A FS $F = \{(x, \mu_F(x)) \mid x \in X\}$ is **normal** if hgt(F) = 1 and **subnormal** if hgt(F) < 1. A FS F is **(identical) null** if $\mu_F(x) = 0 \ \forall x \in X$ and **universal** if $\mu_F(x) = 1 \ \forall x \in X$.

A special emphasis in fuzzy control is given to the **fuzzy congruence** of FSs. For the sake of presenting the contents of this feature the notion of α-height cut in an FS will be first presented. Let $F = \{(x, \mu_F(x)) \mid x \in X\}$ be a FS and α a real number, $\alpha \in (0, 1]$. The α-**height cut** in the FS F, with the notation F_α, is the FS

$$F_\alpha = \{(x, \mu_{F\alpha}(x)) \mid x \in X\}, \quad \mu_{F\alpha}(x) = \begin{cases} 1, & \text{if } \mu_F(x) \geq \alpha, \\ 0, & \text{otherwise,} \end{cases} \quad x \in X, \tag{1.15}$$

and it can be also considered as a crisp set since $\mu_{F\alpha}$ can be viewed as a characteristic function.

Two FSs $F_1 = \{(x, \mu_{F1}(x)) \mid x \in X\}$ and $F_2 = \{(x, \mu_{F2}(x)) \mid x \in X\}$ are **fuzzy congruent** if for any height α, $\alpha \in (0, 1]$, two real numbers α_1, α_2 exist, with the feature $0 < \alpha_1, \alpha_2 \leq 1$, such that

$$\text{Supp}\big((\alpha_1 F_1)_\alpha\big) \subseteq \text{Supp}\big((F_2)_\alpha\big), \ \text{Supp}\big((\alpha_2 F_2)_\alpha\big) \subseteq \text{Supp}\big((F_1)_\alpha\big), \tag{1.16}$$

where the multiplication stands for the multiplication of the m.f.s.

It has been proved by Preitl and Precup (1997) that the fuzzy congruence is equivalent to equal supports. A constraint concerning the fuzzy congruence of two FSs is given if the same support and the same core are imposed on the two FSs, which is equivalent to the **strictly fuzzy congruence**.

The fuzzy congruence and the strictly fuzzy congruence have two *consequences*, 1 and 2, which are remarkable from the point of view of fuzzy control:

1. The monotonous increase/decrease of the flanks of the m.f.s (for certain m.f.s) do not represent a requirement for the strictly fuzzy congruence of

FSs. Therefore, in practical control applications, where the fuzzy information must be processed as quickly as possible, the flanks of m.f.s are built from straight line segments, that is, the m.f.s will be of chosen classical shapes, that is, triangular, rectangular, trapezoidal, or singleton.

2. Major modifications in the characterization of fuzzy information are mainly obtained by modifying the support and/or core of FSs.

The equality and inclusion of two FSs are derived on the basis of crisp set theory. Two FSs $A = \{(x, \mu_A(x)) | x \in X\}$ and $B = \{(x, \mu_B(x)) | x \in X\}$ are **equal**, with the nomenclature $A = B$, if each element of the universe has the same membership degree in both sets, that is,

$$\mu_A(x) = \mu_B(x) \quad \forall x \in X. \tag{1.17}$$

A FS $A = \{(x, \mu_A(x)) | x \in X\}$ is named **fuzzy subset** of the FS $B = \{(x, \mu_B(x)) | x \in X\}$, with the notation $A \subseteq B$, if

$$\mu_A(x) \leq \mu_B(x) \quad \forall x \in X. \tag{1.18}$$

The *set-theoretic operators* (*operators on FSs*) intersection, union, and complement are employed to connect the fuzzy propositions using the logical operators AND, OR, and NOT, respectively, similar to the case of crisp sets. The following operators are frequently used considering again the two FSs $A = \{(x, \mu_A(x)) | x \in X\}$ and $B = \{(x, \mu_B(x)) | x \in X\}$:

−   The **MIN** operator for **intersection**, $A \cap B$:

$$A \cap B = \{(x, \mu_{A \cap B}(x)) | x \in X\}, \quad \mu_{A \cap B}(x) = \min(\mu_A(x), \mu_B(x)) \quad \forall x \in X, \tag{1.19}$$

−   The **MAX** operator for **union**, $A \cup B$:

$$A \cup B = \{(x, \mu_{A \cup B}(x)) | x \in X\}, \quad \mu_{A \cup B}(x) = \max(\mu_A(x), \mu_B(x)) \quad \forall x \in X, \tag{1.20}$$

−   **Mamdani's fuzzy complement** operator for **complement**, A^c:

$$A^c = \{(x, \mu_{A_c}(x)) | x \in X\}, \quad \mu_{A_c}(x) = 1 - \mu_A(x) \quad \forall x \in X. \tag{1.21}$$

However, as a general rule, the *triangular norms* (*t-norms*), *triangular conorms* (*s-norms* or *t-conorms*), and *c-norms* (Klement et al., 2000) are defined to represent the intersection, union, and complement, respectively. A representative t-norm example is the PROD operator:

$$\begin{aligned} \mathrm{PROD}(A, B) &= \left\{ \left(x, \mu_{\mathrm{PROD}(A, B)}(x) \right) | x \in X \right\}, \quad \mu_{\mathrm{PROD}(A, B)}(x) \\ &= \mu_A(x) \mu_B(x) \quad \forall x \in X, \end{aligned} \tag{1.22}$$

and a representative example of s-norm is the SUM operator:

$$\text{SUM}(A, B) = \left\{ \left(x, \mu_{\text{SUM}(A, B)}(x) \right) \mid x \in X \right\}, \quad \mu_{\text{SUM}(A, B)}(x)$$
$$= (\mu_A(x) + \mu_B(x))/2 \quad \forall x \in X, \tag{1.23}$$

where the arithmetic mean must be taken into account in the case of more FSs in order to avoid membership degrees greater than 1. The parameterized operators fuzzy AND and fuzzy OR result in the FS $F = \{(x, \mu_F(x)) \mid x \in X\}$, with the m.f.s defined as follows for fuzzy AND:

$$\mu_F(x) = \gamma \min (\mu_A(x), \mu_B(x)) + (1 - \gamma)(\mu_A(x) + \mu_B(x))/2 \quad \forall x \in X, \tag{1.24}$$

and fuzzy OR:

$$\mu_F(x) = \gamma \max (\mu_A(x), \mu_B(x)) + (1 - \gamma)(\mu_A(x) + \mu_B(x))/2 \quad \forall x \in X, \tag{1.25}$$

and the parameter γ, $\gamma \in [0, 1]$ ensures the possibility to apply these operators in order to represent the intersection or union. A similar possibility is also ensured in the case of the MIN–MAX operator, which is however rarely used in fuzzy control:

$$\mu_F(x) = \gamma \min (\mu_A(x), \mu_B(x)) + (1 - \gamma) \max (\mu_A(x), \mu_B(x)) \quad \forall x \in X. \tag{1.26}$$

Several well-accepted t-norms according to Hamacher, Frank, Yager, and Dubois-Prade, s-norms according to Sugeno, and c-norms according to Sugeno and Yager are defined by Zimmermann (1991) and Driankov et al. (1993). The relaxation of the conditions of these operators leads to other categories of operators, which are used in the mechanisms of fuzzy controllers as uninorms, nullnorms, entropy-based operators, and several metrics (Rudas and Kaynak, 1998; Rudas and Fodor, 2006).

The *modification operators of FSs*, also called *modifiers*, represent operators based on arithmetical computations on the (m.f.s of) FSs and are mainly meant for modeling linguistic hedges as "more/less...," "relatively more/less...," "very...." The particular feature of these operators is that they do not affect the support and core of the FSs, hence they ensure the strict congruence of FSs. The frequently used modifiers and their application to the initial FS $A = \{(x, \mu_A(x)) \,|\, x \in X\}$ are as follows:

- the **concentration** operator, CON(A), with the result in a "denser" FS than the initial one:

$$\mathrm{CON}(A) = \left\{ \left(x, \mu_{\mathrm{CON}(A)}(x) \right) \,|\, x \in X \right\}, \quad \mu_{\mathrm{CON}(A)}(x) = (\mu_A(x))^n,$$
$$n = 2, 3, \ldots, \quad \forall x \in X, \tag{1.27}$$

- the **dilation** operator, DIL(A), with the result in a "less dense" FS in comparison with the initial one:

$$\mathrm{DIL}(A) = \left\{ \left(x, \mu_{\mathrm{DIL}(A)}(x) \right) \,|\, x \in X \right\}, \quad \mu_{\mathrm{DIL}(A)}(x) = \sqrt[n]{\mu_A(x)},$$
$$n = 2, 3, \ldots, \quad \forall x \in X, \tag{1.28}$$

- the **contrast intensification** operator, INT(A), with the result in an FS "with intensified contrast" with respect to the original FS

$$\mathrm{INT}(A) = \left\{ \left(x, \mu_{\mathrm{INT}(A)}(x) \right) \,|\, x \in X \right\},$$
$$\mu_{\mathrm{INT}(A)}(x) = \begin{cases} 2(\mu_A(x))^2 & \text{if } \mu_A(x) < 1/2, \\ 1 - 2\left(1 - (\mu_A(x))^2\right) & \text{otherwise,} \end{cases} \quad \forall x \in X. \tag{1.29}$$

Fuzzy relations represent the basis for the implications in fuzzy logic. Considering the universes $X_1, X_2, \ldots, X_n$, the **n-ary fuzzy relation** is a FS R on the Cartesian product $X_1 \times X_2 \times \cdots \times X_n$.

$$R = \{((x_1, x_2, \ldots, x_n), \mu_R(x_1, x_2, \ldots, x_n)) \,|\, (x_1, x_2, \ldots, x_n) \in X_1 \times X_2 \times \cdots \times X_n\},$$
$$\mu_R : X_1 \times X_2 \times \cdots \times X_n \to [0, 1]. \tag{1.30}$$

Since fuzzy relations are FSs, the operators mentioned before are applied to fuzzy relations as well. In addition, the composition, denoted by $\circ$, is also defined to connect sets and fuzzy relations. Let the FS $A = \{(x, \mu_A(x)) \,|\, x \in X\}$ and the fuzzy relation $R = \{((x, y), \mu_R(x, y)) \,|\, (x, y) \in X \times Y\}$. Then the **MAX-MIN composition (product)** leads to the FS B.

$$B = A \circ R = \{(y, \mu_B(y)) \,|\, y \in Y\}, \quad \mu_B(y) = \max_{x \in X} \min (\mu_A(x), \mu_R(x, y))$$
$$\forall y \in Y. \tag{1.31}$$

The calculations specific to Eq. (1.31) in the case of fuzzy relations defined on countable or discrete universes are done by expressing the operands in matrix forms and applying the matrix computation rules, where the classical product of elements is carried out using the MIN operator, and the sum of products in the classical case (it is actually the result of applying the MIN operator) is carried out using the MAX operator. Other composition versions often used in fuzzy control are

- The **MAX-PROD composition**, where the classical product of elements is carried out using the PROD operator and the sum of products is carried out using the MAX operator.
- The **SUM-PROD composition**, where the classical product of elements is carried out using the PROD operator and the sum of products is carried out using the SUM operator.

1.3 Information processing in fuzzy controllers

This section is dedicated to the presentation of the operating mechanisms in the structure of an FC according to the structure given in Fig. 1.1.

1.3.1 The fuzzification module

The fundamental knowledge representation unit in fuzzy information processing is the notion of *linguistic variable* (**LV**), associated to the structure (N_X, T_X, D_X, M_X), where:

- N_X—the symbolic name of an LV, for example, N_X=speed, temperature, distance, control error, derivative/increment of control error, control signal, etc.
- T_X—the set of *linguistic terms/values* (**LTs**), which represent the linguistic values that can take N_X. An LT associated to an LV, denoted by LT_X as an element that belongs to the set T_X, is a symbol for a particular feature of N_X. In order to define an LV and the corresponding LTs as an essential part in the fuzzification module of an FC, Fig. 1.2 illustrates the transformation of the crisp value of furnace temperature θ_f (LV) into a fuzzy representation. The example deals with the fuzzification of an input variable. For example, the following linguistic terms may be defined in the case of temperature LV ($N_X = \theta_f$): $T_X = \{$VST (very small temperature), ST (small temperature), AT (average temperature), BT (big temperature), VBT (very big temperature)$\}$.

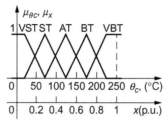

Fig. 1.2 Example to illustrate the fuzzification of an input variable.

- D_X—the domain of crisp values of the LV N_X. In the case of the LV θ_f it is obtained that $D_{\theta_f} = [0, 250]°\,\text{C}$. D_X can be also the universe associated to a FS. The example presented in Fig. 1.2 highlights both the need to define the LTs employed in the fuzzy characterization of crisp information, and the normalization (transformation into normalized/per unit (p.u.) values) of the universe by dividing to the maximum value, $\theta_{fmax} = 250°\,\text{C}$, leading to $x = \theta_c/\theta_{cmax}$ expressed in p.u.. The result is $D_x = [0, 1]$p. u. for the normalized domain.
- M_X—the semantic function, which gives a meaning of an LT in terms of the elements of D_X.

$$M_X : LT_X \rightarrow \overline{LT}_X, \tag{1.32}$$

where $\overline{LT}_X$ is a FS on D_X:

$$\overline{LT}_X = \{(x, \mu_{\overline{LT}_X}(x)) \mid x \in D_X\}, \ \mu_{\overline{LT}_X} : D_X \rightarrow [0, 1]. \tag{1.33}$$

In other words, the function M_X maps a linguistic term/a symbol to an interpretation expressed as a FS. This is the way to differentiate between LTs regarded as either symbols or FSs.

Remark For the sake of simplifying the presentation, the linguistic variables will be next denoted identically by the physical variables they correspond, and their meaning will result from the context.

The crisp information concerning the evolution of CP must be subjected to the following transforms representing **the steps of the fuzzification module** in order to be further processed by the inference module:
- Analog-to-digital conversion of crisp information, that is, sampling the analog signals then quantizing the sampled signals.
- Digital processing of sampled and quantized crisp information, processing of measured signals in terms of digital filtering, and possibly also differentiation and integration.

— Processing the crisp information in a fuzzy expression in terms of LVs and corresponding LTS. Accepting that the FC inputs have well-stated values, for the fuzzy characterization of crisp information it is necessary to define the number of LTs and their m.f.s for each input LV.

The parameters in the fuzzification module to be chosen by the designer are the m.f.s of the LTs corresponding to the input LVs and eventually the scaling factors to be mentioned as follows. The literature does not provide general-purpose exhaustive recommendations with this regard, the final solution representing designer's option. Some relatively general recommendations in this context will be presented here aiming their application to fuzzy controllers design.

1. **Recommendations to choose the number of LTs corresponding to an LV**. This number is usually set to 3, 5, or 7. The number of LTs sets the resolution of further information processing in the FC. On the basis of several case studies, the literature proves—excepting certain special applications that an increase in the number of LTs over seven does not contribute significantly to the efficient resolution increase. However, once this number increases for each input LV, it will lead to an increased number of fuzzy rules and the formulation of the rule base will become more and more difficult. The LTs are generally names to reflect as general as possible the contents and they depend always on the variable involved.

2. **Recommendations to define and use the universe of input (linguistic) variables**. The universe of input variables is predefined by the variation domain of sensors and interfaces (adaptation and conversion). Covering by LTs this domain will determine (in correlation with actuator's properties) the gain of the FC. The existence of several inputs determines the possibility to define around a steady-state operating point more gains, one for each input. The reasoning and corresponding mathematical characterization are the same as in the case of conventional control and are similar to the definition of the *proportional band* of a conventional controller.

The universe can be defined in several ways accounting for the nature of variables involved, the most frequent ways to express it being:
— in natural units,
— in p.u., in terms of division by a (nominal or maximum) value belonging to the universe (Fig. 1.2),
— in increments with respect to a reference value, expressed in natural units or p.u.

The definition of the universe is also referred to as *scaling*, and in the case of using p.u. it is called **input normalization**. However, this operation must be seen in correlation with the universe of output (linguistic) variables, which requires the **output denormalization**. Both operations are necessary in the case of discrete and continuous universes as well. Besides the normalization and denormalization in terms of multiplying/dividing by **scaling factors** that involve nominal or maximum values belonging to the universe, other values of scaling factors can also be used in either linear or nonlinear normalization and/or denormalization. The scaling factors represent key parameters of the FC. So, the choice of their values is important because they affect the gains of the FCs and, further on, the dynamic performance indices of FCS resulted after design. Furthermore, they may represent sources of instability and oscillations (Driankov et al., 1993; Passino and Yurkovich, 1998).

3. **Recommendations for the initial choice of the m.f.s of LTs corresponding to the input LVs**. In the cases of no available application-oriented experience in defining the LTs and m.f.s in case of input LVs (experience gained by case studies and implementations), the following recommendations should be fulfilled in the first phase of design (Precup and Preitl, 1999):

   – The m.f.s of the LTs corresponding to the input LVs are of chosen triangular or trapezoidal type; the m.f.s have (if possible) symmetrical shape excepting the extremity m.f.s.

   – Those allocations of m.f.s are preferred that ensure the total overlap/covering of the universe, so each crisp value should simultaneously fire two LTs (finally, two rules). The overlap of the universe by a single LT could cause discontinuities in the input–output static map of the FC. The intersection point of the m.f.s for two adjacent (with overlap) LTs is recommended to have the ordinate greater than 0.4 (0.45), excepting the extremity zones of the universe, but in the case of LTs corresponding to an output LV this recommendation has to be correlated with the used defuzzification method to be presented later.

   – If the LV involved varies with $\pm$ values around zero, as it is, for example, in the situation of control error, symmetrical allocation with respect to zero is preferred.

   – Zones of the universe that have no overlap by LTs are not accepted. No overlap creates uncertainties in the crisp control signal u, having as effect, for example, $u = 0$.

- The definition of LTs that simultaneously have common subdomains of their cores is not accepted.
- The too rough quantization of the m.f.s results in possible deformations of m.f. support/FS core, with effects on the inference module.

1.3.2 The inference module

According to the structure given in Fig. 1.1, the two fundamental elements of the inference module are the **rule base** and the **inference mechanism** or the **inference engine**, assisted by an adequately defined database. The rule base can be expressed in either symbolic form of type (1.1), or as an inference matrix (or inference table or decision table or MacVicar-Whelan table) that will be exemplified in this section.

Considering the symbolic description of the rule base where a rule is expressed in terms of

$$\text{IF } (E = A) \text{ THEN } (U = B), \tag{1.34}$$

where E and U are input and output linguistic variables, respectively, A and B represent a certain LT of E and U, respectively ($A \in T_E$, $B \in T_U$), the rule interpretation is given by a fuzzy relation on $D_E \times D_U$, with D_E and D_U being the universes of LVs E and U, respectively. The **construction of this fuzzy relation** is carried out in the following steps:

1. The interpretation of the fuzzy proposition $(E = A)$, referred to as **rule antecedent (premise)**, is the FS $\overline{A}$:

$$\overline{A} = \left\{ \left(e, \mu_{\overline{A}}(x) \right) \mid e \in D_E \right\}, \quad \mu_{\overline{A}} : D_E \to [0, 1]. \tag{1.35}$$

2. The interpretation of the fuzzy proposition $(U = B)$, called **rule consequent (conclusion)**, is the FS $\overline{B}$:

$$\overline{B} = \left\{ \left(u, \mu_{\overline{B}}(y) \right) \mid u \in D_U \right\}, \quad \mu_{\overline{B}} : D_U \to [0, 1]. \tag{1.36}$$

3. The interpretation of the rule is given by the fuzzy relation R that ensures the **implication**:

$$R = \left\{ \left((e, u), \mu_R(e, u) \right) \mid (e, u) \in D_E \times D_U \right\}, \quad \mu_R : D_E \times D_U \to [0, 1],$$

$$\mu_R(e, u) = \mu_A(e) * \mu_B(u), \quad \forall (e, u) \in D_E \times D_U, \tag{1.37}$$

where $*$ can be either Cartesian product or any other **fuzzy implication** operator. For example, the MIN operator is used in the case of Mamdani's implication.

Fuzzy propositions as those presented in Eq. (1.34) are simple *fuzzy propositions*. However, in many fuzzy control applications the rule antecedents or consequents are composed fuzzy propositions, that is, simple fuzzy proposition connected by means of intersection, union, and/or complement operators. In this case, before step 1 and eventually before step 2 the inference mechanism should determine the m.f.s corresponding to each such proposition using adequate operators mentioned in Section 1.2.

Subsequently, the transition from fuzzy information processing using a single rule to the case of more rules, the *Mamdani fuzzy rule bases* consist of the rules $R^{(k)}$

$$R^{(k)} : \text{IF } \left(E = A^{(k)} \right) \text{ THEN } \left(U = B^{(k)} \right), \quad k = 1, \dots, n, \tag{1.38}$$

where $A^{(k)}$ is an LT corresponding to the input VL E, $A^{(k)} \in T_E$, the premise interpretation is the FS $\overline{A}^{(k)}$.

$$\overline{A}^{(k)} = \left\{ \left(e, \mu_{\overline{A}(k)}(e) \right) \mid e \in D_E \right\}, \quad \mu_{\overline{A}(k)} : D_E \to [0, 1], \quad k = 1, \dots, n, \tag{1.39}$$

$B^{(k)}$ is an LT corresponding to the output VL U, $B^{(k)} \in T_U$, and the conclusion interpretation is the FS $\overline{B}^{(k)}$.

$$\overline{B}^{(k)} = \left\{ \left(u, \mu_{\overline{B}(k)}(u) \right) \mid u \in D_U \right\}, \quad \mu_{\overline{B}(k)} : D_U \to [0, 1], \quad k = 1, \dots, n. \tag{1.40}$$

Mamdani's implication (characterized by the usage of MIN operator in implication) is interpreted in terms of the fuzzy relation $\overline{R}^{(k)}$.

$$\overline{R}^{(k)} = \left\{ \left((e, u), \mu_{\overline{R}(k)}(e, u) \right) \mid (e, u) \in D_E \times D_U \right\}, \quad \mu_{\overline{R}(k)} : D_E \times D_U \to [0, 1],$$
$$\mu_{\overline{R}(k)}(e, u) = \min \left(\mu_A(e), \mu_B(u) \right), \quad \forall (e, u) \in D_E \times D_U, \quad k = 1, \dots, n, \tag{1.41}$$

where $\mu_{\overline{R}(k)}(e, u)$ indicates the *degree of fulfillment* (the *firing strength*) of the rule $R^{(k)}$.

The interpretation of the rule base is done by the fuzzy relation $\overline{R}$ as a result of the aggregation of all rules in terms of the union of fuzzy relations $\overline{R}^{(k)}$ afferent to the rules $R^{(k)}$.

$$\overline{R} = \bigcup_{k=1}^{n} \overline{R}^{(k)}. \tag{1.42}$$

Employing the MAX operator for the union used in *rule aggregation* (however, the term "aggregation" is also used for connecting simple fuzzy propositions as part of composed fuzzy propositions), Eqs. (1.41) and (1.42) result in

$$\overline{R} = \left\{ \left((e, u), \mu_{\overline{R}}(e, u)\right) \mid (e, u) \in D_E \times D_U \right\}, \quad \mu_{\overline{R}} : D_E \times D_U \rightarrow [0, 1],$$

$$\mu_{\overline{R}}(e, u) = \max_{k=1,\ldots,n} \min \left(\mu_{A(k)}(e), \mu_{B(k)}(u) \right), \quad \forall (e, u) \in D_E \times D_U.$$

$$(1.43)$$

In the final part of the inference module, accepting that the input variable e takes the crisp value e^*, $e = e^*$, the result of applying the rule base (1.38) is interpreted as the FS $\overline{U}$ that represents the *fuzzy conclusion* (*control signal*):

$$\overline{U} = \left\{ \left(u, \mu_{\overline{U}}(u)\right) \mid u \in D_U \right\}, \quad \mu_{\overline{U}} : D_U \rightarrow [0, 1],$$

$$\mu_{\overline{U}}(u) = \max_{k=1,\ldots,n} \min \left(\mu_{A(k)}(e^*), \mu_{B(k)}(u) \right), \quad \forall u \in D_U.$$

$$(1.44)$$

The interpretation of the rule base presented before corresponds to Mamdani's MAX–MIN composition (also referred to as the MAX–MIN inference mechanism or Mamdani's MAX–MIN compositional rule of inference) and characterizes Mamdani fuzzy controllers/fuzzy inference systems. Concluding, this inference mechanism is characterized by: the treatment of AND linguistic connectors in the premise (the intersection of simple fuzzy propositions in the composed fuzzy proposition as part of the premise) by the MIN operator, the treatment of OR linguistic connectors in the premise (the union of simple fuzzy propositions in the composed fuzzy proposition as part of the premise) by the MAX operator, the implication using the MIN operator, and the rule aggregation in terms of the MAX operator.

Other inference mechanisms, which are representative of fuzzy control, are the MAX-PROD and SUM-PROD inference mechanisms. The **MAX-PROD inference mechanism** is characterized by: the treatment of AND linguistic connectors in the premise (the intersection of simple fuzzy propositions in the composed fuzzy proposition as part of the premise) by the MIN operator, the treatment of OR linguistic connectors in the premise (the union of simple fuzzy propositions in the composed fuzzy proposition as part of the premise) by the MAX operator, the implication using the PROD operator, and the rule aggregation in terms of the MAX operator. The **SUM-PROD inference mechanism** is characterized by: the treatment of AND linguistic connectors in the premise (the intersection of simple fuzzy propositions in the composed fuzzy proposition as part of the premise) by the PROD operator, the treatment of OR linguistic connectors in the premise (the union of simple fuzzy propositions in the composed fuzzy proposition as part of the premise) by the SUM operator, the implication using the PROD operator, and the rule aggregation in terms of the SUM operator.

The following example illustrates the operating principle of the MAX-MIN inference mechanism. The example deals with the automatic braking of a train that approaches red lighted traffic lights, which finally implies the train is stopping. The available information for decision-making in braking is arranged under two input variables: d–the distance between the train and the traffic lights, constrained to $d \leq 1000$ m, and v–the train velocity (speed), constrained to $v \leq 100$ km/h.

It is assumed that the railroad and the railway engine have sensors, which can provide with sufficient accuracy the input variables. The introduction of additional conditionings could be taken into consideration in terms of two situations: (i) the definition of additional LVs (e.g., the train mass could belong to this category), (ii) the suitable adaptation of the m.f.s of the LTs corresponding to the input and output LVs.

For the sake of simplicity, five LTs are assigned to each input LV, with the m.f.s defined in accordance with Fig. 1.3, for d: VS, S, A, B, VB, for v: VS, S, A, B, VB, and the nomenclature used are VS–very small, S–small, A–average, B–big, and VB–very big. The output considered in this example is the braking force, f, expressed as degree of train braking in normalized values, f (p. u.), $f \leq f_{max} = 1$, and five LTs with the same symbols as those for the inputs are used for the output as well (Fig. 1.3). The situations with d and v outside the universes are not analyzed. The braking system of the train can be accounted for the choice of the defuzzification method.

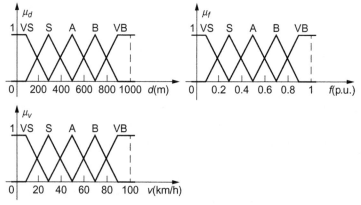

Fig. 1.3 Input and output m.f.s for train braking example.

The chosen shape for the m.f.s of different LTs is trapezoidal for the extremity LTs and symmetrical triangular for the middle LTs. The entire universes of distance and speed are covered and mainly overlapped by LTs, with the usual situation characterized by firing two LTs (for both d and v) and, finally, four rules. The rule base is expressed as the decision table presented in Table 1.1.

The rule base can be defined by an expert on the basis of his/her own experience and can be more or less extended because of some extreme situations in the antecedent that may result in the same consequent. The rule base accepted here is complete (this is obviously an initial situation), consisting of 25 rules (5 LTs corresponding to d) × (5 LTs corresponding to v), and determines the selective firing of any of the five output LTs. The correlation antecedent → consequent for each rule and the entire rule base are illustrated in Table 1.1 as inference table/decision table.

The firing of one or another rule depends on the current (crisp) values of input variables, d^* and v^*. The operating principle of the MAX-MIN inference mechanism is illustrated in a schematic form in Fig. 1.4 for the crisp values of current inputs $d^* = 350$ m and $v^* = 92$ km/h. Fig. 1.4 shows that this case leads to the firing of two LTs for the LV d (namely, S and A) and one LT for the LV v (namely, VB). Therefore, only the following two rules in the rule base will be fired, and the superscripts are obtained according to Table 1.1:

$$R^{(25)} : \text{IF } (d = S \text{ AND } v = VB) \text{ THEN } (f = VB) \text{ OR}$$
$$R^{(35)} : \text{IF } (d = A \text{ AND } v = VB) \text{ THEN } (f = B). \tag{1.45}$$

The value of v^* was deliberately chosen as $v^* > 90$ km/h in order to decrease the number of fired rules and thus to enable a relatively easily understandable presentation of the inference engine. Nevertheless, the number of fired rules would be four in the case of the chosen value

Table 1.1 Rule base expressed as inference table for train braking example

f		v					
		VS	S	A	B	VB	
d	VS	A	A	B	VB	VB	1
	S	S	A	B	V	VB	2
	A	VS	S	A	B	B	3
	B	VS	S	S	A	B	4
	VB	VS	VS	VS	S	A	5
		1	2	3	4	5	

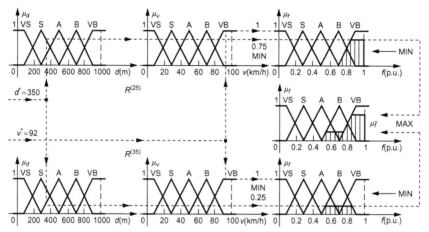

Fig. 1.4 MAX-MIN inference mechanism for train braking example.

$0 < v^* < 90$ km/h. Finally, the fuzzy control signal is expressed as a FS with the m.f. $\mu_{\bar{f}}$ according to Fig. 1.4.

The inference mechanisms together with the rules presented before employ Mamdani fuzzy control rules, applicable mainly to fuzzy control with Mamdani fuzzy controllers. The **Takagi–Sugeno fuzzy rules** (Takagi and Sugeno, 1985) are applicable in both fuzzy control–on the basis of **Takagi–Sugeno fuzzy controllers**, referred to also as **Takagi–Sugeno–Kang fuzzy controllers**–and systems modeling in the case of nonlinear dynamical processes with behaviors subjected to characterization in terms of a set of operating regimes (defined in the vicinity of a set of operating points). The most general form of these rule bases is

$$R^{(k)} : \text{IF } \left(E = A^{(k)}\right) \text{ THEN } (U = f_k(E)), \quad k = 1, \ldots, n, \qquad (1.46)$$

where the premise part is identical to Mamdani's case and the difference appears in the conclusion by the fact that $f_k(E)$ is a nonlinear or linear function that describes the dynamics of the process/fuzzy controller or even the dynamics of the FCS for the particular value of the LT $A^{(k)}$ that corresponds to the input LV E.

The **inference mechanisms** used in the case of **Takagi–Sugeno** rule bases are similar to those in Mamdani's case, with the obvious difference in the rule aggregation part due to the different expressions in rule conclusions. Accepting an input variable e taking the crisp value e^*, $e = e^*$, this special form of the conclusions results in the fuzzy conclusion expressed as the (fuzzy) set of singletons.

$$\left\{ \left(u_{(k)}, \mu_{(k)} = \mu_{\overline{R}(k)} \left(e^*, u_{(k)} \right) \right) \middle| \left(e^*, u_{(k)} \right) \in D_E \times D_U \right\}, \quad k = 1, \dots, n,$$

(1.47)

where $u_{(k)}$ is the modal value afferent to each singleton (i.e., to each fired rule) and $\mu_{(k)} = \mu_{\overline{R}(k)} \left(e^*, u_{(k)} \right)$ is the firing strength of rule $R^{(k)}$.

If the LV VL E corresponds to the state vector, then $A^{(k)}$ defines a fuzzy subset of the FS on the state space corresponding to a particular operating regime and $f_k(E)$ describes the dynamics of process/system in this operating regime. This type of fuzzy rules (associated to a certain inference mechanism) can be regarded as an interpolating mechanism that weights more or less certain local models/controllers afferent to different operating regimes depending on the current operating point.

The parameters in the inference module to be set by the designer are the rule base and the inference mechanism. The **rule base** should be defined correctly to ensure the fulfillment of the following three important **properties** that ensure a good operation of the FC, 1, 2, and 3:

1. A **rule base** must be **complete**, that is, any combination of input values leads to a certain output value. In relation to Eq. (1.45), this means that.

$$\text{hgt}(\overline{U}) > 0 \quad \forall e^* \in D_E.$$

(1.48)

The number of rules as part of a complete rule base (illustrated in the train braking example) is calculated as the product of numbers of LTs that correspond to the input LVs.

2. A rule base must be **consistent**, that is, it should not contain contradictions. In other words, it should not contain rules with the same rule antecedent but having mutually exclusive rule consequents (the same premise always leads to the same conclusion).

3. A rule base must be **continuous**, that is, it should not contain neighboring rules with output FSs that have empty intersection. Two rules are considered neighboring if the intersection of the FSs, obtained as a result of their premises (of type (1.40)), is a FS with a nonzero height (Driankov, 2001).

The **definition of rule base** becomes more and more difficult if the number of inputs increases. General recommendations to define the rule base cannot be given; the role of expert (who knows the evolution of the process) is of exquisite importance. The following **methods** are recommended by the literature as suitable for the definition of rule bases:

- The use of an expert's knowledge in controlling the given process, an adequately prepared interview being used to formulate expert's knowledge.
- Heuristic methods, based on engineering-like analyses of the possible evolutions of the process variables, done in cooperation with the process engineer.
- The use of past experimental results, real-world or simulated ones, in situations when more or less detailed mathematical models of the CP are available.
- Special methods employing knowledge gained from conventional controllers and identification techniques that involve clustering, (neural) network-based learning, assisted by optimization algorithms including nature-inspired ones.

As far as the **choice of inference mechanism** is concerned, the usual choice deals with one of the three mechanisms previously discussed, viz. MAX-MIN, MAX-PROD, and SUM-PROD. The case studies investigated by several authors show that the inference mechanism has less significant effect on the shape of FC input-output map. Therefore, it is preferred to work with the methods that ensure good efficiency in information processing.

1.3.3 The defuzzification module

The defuzzification is the conversion of the fuzzy control signal, which is a FS as result of the inference module, into a crisp value. The defuzzification is necessary because the actuators need crisp signals that can be interpreted in contrast with FSs. It is obvious that the crisp value calculated by defuzzification should belong to the universe of the fuzzy control signal. In case of using a normalized domain for the control signal/output of the FC, the *defuzzification module* consists of two sub-modules: the strictly speaking defuzzification one explained above, and the *output denormalization*, which maps the crisp value of output onto its physical domain.

A lot of defuzzification methods are used in the case of *Mamdani fuzzy controllers*. The two most often used defuzzification methods for these fuzzy controllers are the mean of maxima (MoM) method and the center of gravity (CoG) method (expressed in several versions in the literature including the version focused on singleton m.f.s of the LTs that correspond to the output LVs).

The **MoM** *method* calculates the average of crisp values of output (control signal) that correspond to the conclusions with maximum firing

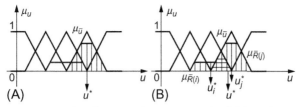

Fig. 1.5 Operating principle of mean of maxima method (A) and center of gravity method (B).

strength. Accepting that the result of inference module is obtained as the FS $\bar{u}$ (on the universe D_u), which represents the fuzzy control signal, the crisp control signal u^* is obtained in terms of

$$u^* = 0.5 \left[\inf_{u \in D_u} \{u \in D_u | \mu_{\bar{u}}(u) = \mathrm{hgt}(\bar{u})\} + \sup_{u \in D_u} \{u \in D_u | \mu_{\bar{u}}(u) = \mathrm{hgt}(\bar{u})\} \right]$$

$$(1.49)$$

and exemplified in Fig. 1.5A.

The **CoG method** determines the crisp value of output taking into consideration, in a weighted manner, all influences obtained from the rules fired by the particular state of the inputs at a certain moment. The formulae giving the crisp control signal are adapted from mechanics and are specific to calculating the abscissa of CoG:

− in the continuous case:

$$u^* = \left[\int_{D_u} u \cdot \mu_{\bar{u}}(u) \mathrm{d}u \right] \bigg/ \left[\int_{D_u} \mu_{\bar{u}}(u) \mathrm{d}u \right], \qquad (1.50)$$

− in the discrete case:

$$u^* = \left[\sum_{i=1}^{m} u_i \cdot \mu_{\bar{u}}(u_i) \right] \bigg/ \left[\sum_{i=1}^{m} \mu_{\bar{u}}(u_i) \right], \quad m = \mathrm{card}(D_u). \qquad (1.51)$$

The operating principle of the CoG method is exemplified in Fig. 1.5B. This illustration shows that the overlapping area is not reflected in the above calculations.

The *weighted average method* is usually used in **Takagi–Sugeno fuzzy controllers**. Considering the fuzzy control signal in Eq. (1.46), the crisp control signal u^* is computed as

$$u^* = \left[\sum_{k=1}^{n} u_{(k)} \cdot \mu_{(k)} \right] \bigg/ \left[\sum_{k=1}^{n} \mu_{(k)} \right]. \qquad (1.52)$$

The parameters in the defuzzification module that have to be set by the designer in the case of Mamdani FCs are the m.f.s of the LTs corresponding to the output LV, the defuzzification method, and the conversion of crisp signal. Several analyses must be performed that enable setting all these parameter values, but all analysis should account for the characteristics of the actuators (which are without dynamics or with negligible dynamics, eventually absorbed by the CP), the minimum control system performance indices to be achieved by the FCS, and the version of FC implementation, through hardware or software construction.

Aspects concerning the design of any FC from the point of view of setting the parameters in the defuzzification module are as follows:

1. **Aspects concerning the choice of linguistic terms and m.f.s corresponding to the output/control signal linguistic variable**. The main aspects of interest in this case can be resumed as follows:

    (a) **The number of chosen LTs** is usually odd (3, 5, or possibly 7). A larger number of LTs brings no spectacular results in the shape of the input-output static map.

    (b) **The existence of zones in the universe with no covering by LTs/m.f.s** does not represent a serious problem. The covering of universe by continuous or discrete crisp values of the control signal is solved using the convenient choice of defuzzification method.

    (c) **The scaling/definition of the universe**. The universe has to be always scaled/defined in such a way that it should fulfill the control requirements of actuator, and the output denormalization related to the fuzzification module should be also taken into consideration. This means that:

    – The crisp control signal is not permitted to exceed the extreme values accepted by the actuator, this exceedance is related to: turning off/on the actuator, dynamic forcing to obtain as reduced as possible settling time.

    – The variation domain of the signal fed to the actuator and, accordingly, the variation domain of the actuating signal must be overlapped sufficiently by the variation domain of control signal in order to reach the operating points imposed to the control system.

    The last remark can also become a specific aspect known as **the extremity problem** of the FC. The incorrect definition of the m.f.s of output LTs in correlation with an inadequate chosen defuzzification method can result in the situation when there is no overlap of the necessary domain by the obtained crisp control signal.

The extremity problem can be avoided in terms of the following measures:

- the symmetrical extension the extremity m.f.s,
- the modification of the shapes of m.f.s, for example, to singleton m.f.s,
- the choice of another defuzzification method (overall or just in the extremity zones).

(d) **The shapes of m.f.s** will be chosen to ensure–in correlation with the defuzzification method–a good efficiency of information processing (usually reflected in as reduced as possible computation time). Singleton, rectangular, triangular, and trapezoidal m.f.s are recommended in this regard, with the following highlights:

- the singleton m.f.s are the easiest ones to be processed,
- the rectangular m.f.s significantly modify the computational cost, but, by varying the width of the rectangle (the support), additional modifications of FC features are obtained, nevertheless, the overlap problem can be avoided in comparison with the situation of triangular m.f.s (the computation of the CoG becomes heavier),
- on the basis of all aspects presented before, the triangular m.f.s seem to be the most unfavorable ones.

2. **Aspects concerning the choice of defuzzification method**. The following criteria have to be accounted for when choosing the defuzzification method (Driankov et al., 1993; Precup and Preitl, 1999):

- The type of actuator. The MoM method is recommended in the case of actuators with finite numbers of discrete states. The CoG method is preferred in the case of actuators with compact variation domains/universes.
- The continuity of the input-output static of FC has to be ensured. This means that a small change in the fuzzy control signal should not result in a large change in the crisp control signal (a small sensitivity). The CoG method satisfies this criterion and the MoM method does not.
- The ambiguity should be avoided. This is related to avoiding situations when two relatively large areas in the m.f.s of fuzzy control signal are covered by two areas in the m.f.s of FSs as result of implication. Both CoG and MoM methods satisfy this criterion.

- The plausibility is necessary. This is characterized by placing the crisp control signal approximately in the middle of the support of the fuzzy control signal. The CoG method does not satisfy this criterion, and the MoM method satisfies this criterion only in combination with the MAX-PROD inference mechanism.
- The computational cost is particularly important in practical applications of fuzzy controllers. The MoM method is a computationally fast method, whereas the CoG method is much slower. Although the use of CoG method seems to be difficult, choosing particular shapes of m.f.s and well-acknowledged inference methods determines faster information processing, and fuzzy control signals with m.f.s having convenient shapes that enable relatively easy analytical calculations.

3. **Conversion of crisp control signal**. Depending on the defuzzification method and on the type of actuator, the crisp control signal will be (in the case of digital control):
   - the current crisp value of control signal $u(t_d)$ or.
   - the increment of current crisp value with respect to the previous crisp value of control signal, $\Delta u(t_d) = u(t_d) - u(t_d - 1)$, where $t_d \in \mathbf{N}$ is the discrete time index.

In both situations the resulted value is converted into an analog form by a digital-to-analog converter, excepting the situations when the actuator accepts directly as input the binary form of the crisp control signal from the case of continuous process control. The problems and results related to information quantization are the same as in the case of analog-to-digital conversion.

Since this section is mainly focused on Mamdani fuzzy systems/controllers, the following section will be dedicated to a short presentation of the operating mechanisms in Takagi-Sugeno fuzzy models (referred to also as Takagi-Sugeno-Kang fuzzy models) and Tsukamoto fuzzy models.

1.3.4 Takagi-Sugeno fuzzy models and Tsukamoto fuzzy models

Takagi–Sugeno fuzzy models, also known as *Takagi–Sugeno-Kang* (*TSK*) *fuzzy models* or *Sugeno models* (Takagi and Sugeno, 1985; Sugeno and Kang, 1988), have been suggested firstly as an alternative to the development of systematic approaches capable of generating fuzzy rules from a given input-output data set. Considering a two input-single output system, a typical fuzzy rule in a Takagi-Sugeno fuzzy model has the following form, which is similar to the rule given in Eq. (1.46):

$$\text{IF } (x = A \text{ AND } y = B) \text{ THEN } (z = f(x, y)), \qquad (1.53)$$

where A and B are FSs in the premise (antecedent) and $f(x, y)$ is a crisp function in the conclusion (consequent). Usually this function is a polynomial in the input variables x and y, but it can be any linear or nonlinear function as long as it can appropriately describe the output of the model, z, in the fuzzy region specified by the rule antecedent. When this function is a first-order polynomial, the resulting fuzzy model (fuzzy inference system) is referred to as *first-order Sugeno fuzzy model*. If $f(x, y)$ is a constant (in fact, more constants, each one appearing in a certain rule), the fuzzy model is called *zero-order Sugeno fuzzy model*, a special case of Mamdani fuzzy inference system described in this chapter. The rule consequents in zero-order Sugeno fuzzy models are specified by singletons or pre-defuzzified consequents.

As mentioned earlier, the **inference mechanisms** involved in Takagi-Sugeno fuzzy models are similar to those in Mamdani fuzzy models (fuzzy inference systems), with the difference in the rule aggregation part due to the different expressions in rule consequents. This determines the fuzzy conclusion expressed as the (fuzzy) set of singletons of type (1.47). The most widely used **defuzzification method** in the case of Takagi-Sugeno fuzzy models is the weighted area method where the calculation of the crisp control signal is given in Eq. (1.52). The other modules in the Takagi-Sugeno fuzzy model structure are identical to previously presented Mamdani's case. Due to these features the outputs of Takagi-Sugeno fuzzy models are smooth functions of their input variables as long as the neighboring m.f.s in the antecedent have enough overlap whereas the overlap of m.f.s in the consequents of Mamdani fuzzy models does not have a decisive effect on the smoothness (the overlap of the antecedent m.f.s plays the key role in this respect).

The operating principle of the Sugeno fuzzy model is illustrated in Fig. 1.6 for a first-order Sugeno fuzzy model. The t–norm used in the inference engine (highlighted in Fig. 1.6) is usually the MIN or PROD operator.

The parameters w_1 and w_2 stand for the firing strengths of the two rules, and the rule consequents in the two rules, z_1 and z_2, that represent the fuzzy consequent, are expressed as

$$z_1 = p_1 x + q_1 y + r_1, \quad z_2 = p_2 x + q_2 y + r_2. \qquad (1.54)$$

Consequently, the crisp output of the fuzzy model, z, is obtained in terms of the weighted area method of defuzzification

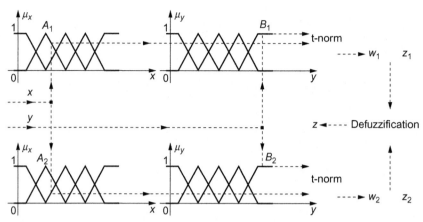

Fig. 1.6 Operating principle of two input-single output first-order Sugeno fuzzy model.

$$z = (w_1 z_1 + w_2 z_2)/(w_1 + w_2). \tag{1.55}$$

Since each rule has a crisp output (z_1 and z_2 in the example considered here), it is often considered in the literature that Takagi-Sugeno fuzzy models do not possess defuzzification modules, the defuzzification being replaced with several operators (Jang et al., 1997). In the case of the weighted area method of defuzzification, this operator (defined in Eq. (1.55)) is known as the *weighted average* operator. In practice, the weighted average operator is sometimes replaced with the *weighted sum* operator in order to reduce the computational costs. This approach is used especially in the case of fuzzy modeling and identification leading to

$$z = w_1 z_1 + w_2 z_2 \tag{1.56}$$

Tsukamoto fuzzy models (Tsukamoto, 1979) are characterized by special rule consequents represented using FSs with monotonically m.f.s. As a result, the inferred output of each rule is defined as a crisp value induced by the rule's firing strength. Then, the crisp output of the fuzzy model, z, is obtained in terms of the weighted area method of defuzzification or, in other words, in terms of taking the weighted average of each rule's output.

The operating principle of the Tsukamoto fuzzy model is illustrated in Fig. 1.7, where the defuzzification is applied according to Eq. (1.55), similar to the case of Takagi-Sugeno fuzzy models. The application of the weighted average method of defuzzification avoids other time-consuming defuzzification methods. However, Tsukamoto fuzzy models are seldom used as they are not as transparent as Mamdani and Takagi-Sugeno models.

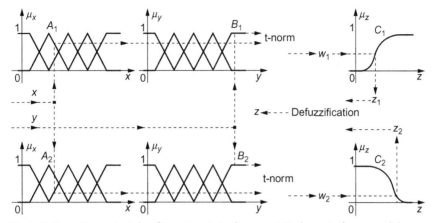

Fig. 1.7 Operating principle of two input-single output Tsukamoto fuzzy model.

Remark Zero-order Sugeno fuzzy models also represent the special cases of Tsukamoto fuzzy models where each rule consequent is specified by a step-type m.f. centered at the constant involved in the rule consequent.

1.4 Fuzzy controllers and design approaches

The *fuzzy controller* (**FC**) represents a nonlinear controller with one or more inputs and one or more outputs. The shape of the nonlinear input-output (static) map attached to the FC can be modeled to achieve a large variety of forms by the adequate setting of the parameters in the modules of the FC. The additional dynamic processing of some of the variables (by differentiation and/or integration) can provide dynamic features that lead to FCs with dynamics. As a result, the entire range of linear or conventional controllers can be generated by extension to fuzzy control. However, it is obvious that all these generated controllers have parameters that are subjected to modifications due to the desired nonlinear input-output maps.

1.4.1 Fuzzy controllers without dynamics

The presentation of the FCs without dynamics will be focused shortly on single input-single output (SISO) nonlinear proportional FCs, SISO two-positional FCs, SISO three-positional FCs, and multi-input-single output (MISO) nonlinear proportional FCs.

1. *The SISO nonlinear proportional FC.* If the input and output universes are covered by a very small number of LTs (one or two LTs), a conventional controller with limitation is obtained. This controller has more or less nonlinear input-output static map depending on the features of the informational modules specific to FC.

(i)   **The FC has a single rule**. Trapezoidal m.f.s are used for the LT of the input and output LV, the universe is symmetrical with respect to 0 in the zone where the membership degrees are less than one, and the CoG defuzzification method is used. The modification of the slope of the input-output static map can be achieved by scaling the universes of the two LTs or modifying the shapes of the two m.f.s.

(ii)   **The FC has two rules**. The two LTs of the input and output LVs make use of trapezoidal m.f.s, the universes are symmetrical with respect to 0 in the zone where the membership degrees are less than one, and the CoG defuzzification method is also used in this case. The modification of the slope of the input-output static map can be achieved as in the case (i).

The further increase in the number of input and output LTs leads to nonlinear input-output static maps with shapes that can be of interest for the designer of FCs to compensate for the nonlinearities of the CP (Preitl and Precup, 1997).

2. *The SISO two-positional and three-positional FCs*. **The two-positional FCs** use two rules and two LTs for the input and output LVs. The two triangular m.f.s of the LT of input and output LVs are symmetrical with respect to 0, and have 0 overlap. Singleton m.f.s are also accepted for the output, but with the model values of the triangular m.f.s. The hysterezis zone is obtained by ensuring a zone with no overlap as far as the LTs of the input LV are concerned.

The three-positional FCs use three rules. The three triangular m.f.s of the LT of input and output LVs are symmetrical with respect to 0, and have again 0 overlap. The other aspects specific to the two-positional FCs are valid for three-positional FCs as well.

The same principle used to allocate the m.f.s of the LTs that correspond to the input LV can lead to multi-positional FCs by increasing the number of LTs. All these FCs use the largest of maxima defuzzification method, which is also referred to as the height method (Precup and Preitl, 1999).

3. *The MISO nonlinear proportional FCs*. The structure of these controllers is shown in Fig. 1.1. The construction of such a controller assumes that the input variables in the input vector $\mathbf{e}'$ and obtained from the CP

have to be signals that are separated by blocks with dynamics inside the CP.

An interesting case for practical applications is that where the input vector $\mathbf{e}'$ is the state vector of the process. This is the way to build **state feedback fuzzy controllers** without or with control error correction (to ensure zero steady-state control error), with measured or estimated state variables.

1.4.2 Fuzzy controllers with dynamics

It is generally acknowledged that the performance of linear or conventional control systems can be enhanced by inserting dynamic components in the controller structure. The effects of these components can be reflected:

— in steady-state regimes, by the rejection or, from one case to another, alleviation of the control error.
— in dynamic regimes, by improving the phase margin (for linear systems), reducing the overshoot, and reducing the settling time, and/or improving (relaxing) the stability conditions.

The same general usefulness can be given by inserting dynamic components in the case of fuzzy controllers as well.

The dynamic processing of an input signal–by creating additional I or D components–creates additional signals considered as FC inputs. If $e_1 = e$ is the control error, then the additional FC inputs e_2 and e_3 can be, for example, $e_2 = \dot{e}$, $e_3 = \ddot{e}$, or $e_3(t) = \int_0^t e(\tau)\mathrm{d}\tau$. As shown by Precup and Preitl (1999), many principles can be used to implement dynamic components, and they lead to several structures with dynamics.

The structures shown in Fig. 1.8 illustrate approaches to the dynamic processing (of D or I type) of different variables that belong to an FC. Both the input signals (prior to defuzzification) and the crisp (defuzzified) output of FC can be subjected to dynamic processing, that is, the dynamic processing is carried out outside the strictly speaking FC, which remains, in essence, nonlinear and without dynamics.

The D or I components can be implemented in either analog or digital version. The symbols of different processing types used in Fig. 1.8 are well known. The analog implementation versions for D components (the processing is of DT1 (derivative with first-order lag) type in the real case) and I components are presented by Bühler (1994), but their importance is reduced in fuzzy control. The digital versions of D (DT1) and I components create quasi-continuous equivalents of analog D and

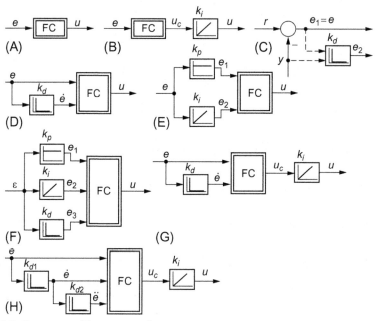

Fig. 1.8 Structures of fuzzy controllers.

I components, respectively. There are several methods for the accomplishment of Q-C D and I components, but only the following are presented:

(a) The usual computation relation for **the D component** $d(t_d)$ is

$$d(t_d) = \frac{e(t_d) - e(t_d - 1)}{T_s}, \quad t_d \in \mathbf{N}^*, \tag{1.57}$$

where T_s is the sampling period. If the input variable e has very rapid variation, which can be harmful to the implementation of the D component, then either e is pre-filtered in terms of a PT1 law, or the D component $d(t_d)$ is created on the basis of the current sample $e(t_d)$ and the past sample $e(t_d - m)$:

$$d(t_d) = \frac{e(t_d) - e(t_d - m)}{m \cdot T_s}, \quad m \in \mathbf{N}, \ m \geq 2. \tag{1.58}$$

The efficiency of Eq. (1.58) should be checked depending on the application involved.

(b) The usual computation relation for **the I component** $\sigma(t_d)$ is

$$\sigma(t_d) = \sum_{i=0}^{t_d} e_i = e(t_d) + \sum_{i=0}^{t_d-1} e_i, \tag{1.59}$$

which is equivalent to

$$\sigma(t_d) = \sigma(t_d - 1) + e(t_d), \quad \sigma(t_d) = x(t_d) + e(t_d), \quad x(t_d) = \sum_{i=0}^{t_d-1} e_i. \quad (1.60)$$

Such a characterization will also allow for a relatively easy quasi-continuous equivalence of the digital case. Using the first-order Padé approximation

$$e^{-s \cdot T_s} \approx \frac{1 - s \cdot T_s/2}{1 + s \cdot T_s/2} \quad (1.61)$$

leads to relationships for the two components with D and I dynamics (Precup and Preitl, 1999)

$$d(s) \approx \frac{s}{1 + s \cdot T_s/2} \cdot e(s), \quad \sigma(s) \approx \frac{1 + s \cdot T_s/2}{s \cdot T_s} \cdot e(s), \quad (1.62)$$

which will contribute to the definition of pseudo-transfer functions attached to FCs with dynamics.

Since FCs are essentially nonlinear, it is obviously wrong to speak about a transfer function (t.f.) of an FC. However, by accepting continuous generalized input-output static maps of the FCs, a quasi-continuous pseudo-transfer function can be assigned to an FC, and it is valid in the vicinity of a steady-state operating point (e.g., the origin of the strictly speaking FC with I component on FC output shown in Fig. 1.8B).

The proportional-integral fuzzy controllers (PI-FCs) are useful as they can be designed systematically starting with the features (that are known or accepted as acceptable) of initial proportional-integral (PI) controllers. In addition, as shown by Åström and Hägglund (1995), PI controllers are widely used, together with proportional-integral-derivative (PID) controllers in the majority of industrial control applications. Moreover, the knowledge on PI controllers with output/control signal integration (Fig. 1.8G) will represent the support to design other FCs synthesized in Fig. 1.8: P fuzzy controller (Fig. 1.8A), I fuzzy controller (Fig. 1.8B), proportional-derivative (PD) fuzzy controller (Fig. 1.8D), PI fuzzy controller with input/control error integration (Fig. 1.8E), PID fuzzy controller with input/control error integration (Fig. 1.8F), and PID fuzzy controller with output/control signal integration (Fig. 1.8H). These are the reasons why **the Mamdani PI-FCs with output/control signal integration** are discussed. The dynamics is inserted in terms of (Fig. 1.9A).

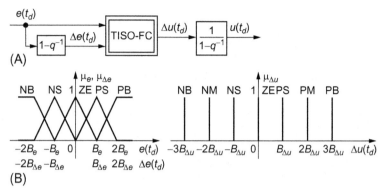

Fig. 1.9 Structure of PI-FC with output integration (A), and input and output membership functions of Mamdani PI-FC (B). *(From David, R.-C., Precup, R.-E., Petriu, E.M., Radac, M.-B., Preitl, S., 2013. Gravitational search algorithm-based design of fuzzy control systems with a reduced parametric sensitivity. Inf. Sci. 247, 154–173.)*

– the numerical differentiation of the control error $e(t_d)$ that leads to the increment of control error $\Delta e(t_d)$.

$$\Delta e(t_d) = e(t_d) - e(t_d - 1), \tag{1.62}$$

– the numerical integration of the increment of control signal $\Delta u(t_d)$.

$$\Delta u(t_d) = u(t_d) - u(t_d - 1). \tag{1.63}$$

The structure and the input m.f.s of a simple Mamdani PI–FC are presented in Fig. 1.9, where q^{-1} is the backward shift operator.

The structure shown in Fig. 1.9A is used for both Mamdani and Takagi-Sugeno PI-FCs, and the nonlinear TISO-FC block indicates the two inputs-single output fuzzy controller, namely the strictly speaking fuzzy controller without dynamics. However, the internal structure of the TISO-FC block is different for Mamdani and Takagi–Sugeno PI-FCs.

The design of this controller starts with the expression of the incremental form of the recurrent equation of a quasi-continuous discrete-time (digital) PI controller.

$$\Delta u(t_d) = K_P \Delta e(t_d) + K_I e(t_d) = K_P[\Delta e(t_d) + \chi \cdot e(t_d)], \tag{1.64}$$

where the parameters K_P, K_I, and χ are obtained from the parameters of the continuous-time (linear) PI controller with the t.f.

$$C(s) = k_C \left(1 + \frac{1}{s \cdot T_i}\right), \tag{1.65}$$

with k_C–the controller gain and T_i–the integral time constant. The application of Tustin's method to discretize the continuous-time PI controller leads to the parameters K_P, K_I, and χ.

$$K_P = k_C \left(1 - \frac{T_s}{2T_i}\right), \quad K_I = k_C \frac{T_s}{T_i}, \quad \chi = \frac{K_I}{K_P} = \frac{2T_s}{2T_i - T_s}. \qquad (1.66)$$

Eq. (1.64) and the representation of the increment of control signal $\Delta u(t_d)$ in the phase plane $\langle e(t_d), \Delta e(t_d)\rangle$ allow for expressing the features that support the design of this fuzzy controller including the formulation of its rule base (Preitl and Precup, 1997):

- there exists a zero control signal line with the equation

$$\Delta e(t_d) + \chi \cdot (t_d) = 0, \qquad (1.67)$$

 and it divides the phase plane in two half-planes,
- $\Delta u(t_d) > 0$ in the upper half-plane and $\Delta u(t_d) < 0$ in the lower half-plane with respect to this line,
- the distance from a point in the phase plane to the zero control signal line corresponds to the modulus of the increment of control signal, $|\Delta u(t_d)|$.

The fuzzification in the Mamdani PI-FC is carried out as follows (Fig. 1.9B):

- for the input variables $e(t_d)$ and $\Delta e(t_d)$: triangular and trapezoidal m.f.s associated to the LTs are set, the m.f.s are uniformly distributed with an overlap of 1,
- for the output variable $\Delta u(t_d)$: singleton m.f.s associated to the LTs are set,
- the scaling factors are set.

The scaling factors have been omitted in Fig. 1.9B for the sake of simplicity. However, they are important as they can scale the universes to convenient intervals and they can also contribute to the modification of the nonlinear input-output map of the fuzzy controller.

Fig. 1.9B highlights the three tuning parameters of the Mamdani PI-FC, namely B_ε, $B_{\Delta e}$, $B_{\Delta u}$, which are related to the shapes of the m.f.s associated to the LTs of input and output LVs. This reduced number of parameters will ensure a cost-effective implementation of the Mamdani PI-FC. A larger number of parameters can be useful for an improved compensation of the nonlinearities specific to the CP.

The rule base of the PI-FC is formulated using the above features. The complete rule base for five LTs of both input LVs is formulated by means of the decision table presented in Table 1.2.

Table 1.2 Rule base expressed as decision table for Mamdani PI-FC with output integration

Δu_k		e_k				
		NB	**NS**	**ZE**	**PS**	**PB**
Δe_k	PB	ZE	PS	PM	PB	PB
	PS	NS	ZE	PS	PM	PB
	ZE	NM	NS	ZE	PS	PM
	NS	NB	NM	NS	ZE	PS
	NB	NB	NB	NM	NS	ZE

The modal equivalence principle (Galichet and Foulloy, 1995) can be applied to design Mamdani fuzzy controllers. This principle states that the initial linear controller and the fuzzy controller designed from it should exhibit the same output values for the inputs equal to their modal values. In this context, the tuning of the parameters B_e, $B_{\Delta e}$, $B_{\Delta u}$ is carried out as follows:

- The following relationship is applied along the zero control signal line:

$$\chi = -\Delta e(t_d)/e(t_d) = B_{\Delta e}/B_e, \tag{1.68}$$

- The following condition is fulfilled along the constant control signal line $\Delta u(t_d) = B_{\Delta u}$:

$$B_{\Delta u} = \Delta u(t_d) = K_P(\Delta e(t_d) + \chi \cdot e(t_d)) = K_P B_{\Delta e}, \tag{1.69}$$

- Eq. (1.69) leads to

$$B_{\Delta u} = K_P \cdot \chi \cdot B_e = K_I \cdot B_e, \tag{1.70}$$

- One of the three parameters (e.g., B_e) is chosen according to the experience of the control system designer, and the other two parameters ($B_{\Delta e} = \alpha \cdot B_e$, $B_{\Delta u} = K_I \cdot B_e$ and $B_{\Delta u}$) are computed in terms of

$$B_{\Delta e} = \chi \cdot B_e, \quad B_{\Delta u} = K_I \cdot B_e. \tag{1.71}$$

The choice of the inference mechanism and the defuzzification method represent designer's option. Mamdani's MAX–MIN inference mechanism and the CoG defuzzification method are usually set.

The incremental form of the control signal, $\Delta u(t_d)$, can be next used in the control system:

- either directly, if the actuator is of integral type or it contains a pure integral component,
- or by expressing the control signal according to

$$u(t_d) = u(t_d - 1) + \Delta u(t_d), \tag{1.72}$$

which corresponds to the numerical integration block placed at the PI-FC output (Fig. 1.9A).

Concluding, the results presented in this section leads to **an approach to design control systems with Mamdani PI-fuzzy controllers**. This approach consists of the following steps:

Step 1. The CP is modeled.

Step 2. A linear PI controller with the t.f. given in Eq. (1.65) is designed and tuned using an approach specific to linear or conventional controllers.

Step 3. The sampling period is set, the PI controller is discretized, and expression (1.64) of the incremental form of the recurrent equation of a quasi-continuous discrete-time (digital) PI controller is obtained.

Step 4. The fuzzy controller structure is set.

Step 5. The value of the tuning parameter B_e is set, and Eq. (1.71), which is a particular expression of the modal equivalence principle, is applied to obtain the values of the other two tuning parameters, $B_{\Delta e}$ and $B_{\Delta u}$.

This design approach can also be applied to Takagi-Sugeno PI-FCs, as shown next in this section. However, only the first equation in Eq. (1.71) is used in the case of Takagi-Sugeno PI-FCs due to the particular structure. The Takagi-Sugeno version proves to be advantageous in several situations due to exploiting the fact that it is a bumpless interpolator between a set of local conventional controllers, in particular PI ones (Babuška and Verbruggen, 1996; Precup and Preitl, 2004). Therefore, **an approach to design control systems with Takagi–Sugeno fuzzy controllers** is expressed using the following design steps:

Step 1. The mathematical modeling of the CP is carried out. Both simplified crisp models and fuzzy ones (Babuška and Verbruggen, 1996; Iglesias et al., 2012; Penedo et al., 2012; Precup et al., 2015) can be used.

Step 2. Local mathematical models of the process are derived. These models depend on the operating conditions specific to the process and the performance imposed to the control system.

Step 3. Local linear or fuzzy controllers are designed to lead to local simplified or local fuzzy models.

Step 4. The fuzzy controller structure is set to ensure the bumpless transfer between the local controllers.

This design approach is known in the literature under different labels, and the most widely used in the case of Takagi-Sugeno fuzzy controllers **is the parallel distributed compensation** (Wang et al., 1995). The application of the modal equivalence principle to Takagi-Sugeno fuzzy controllers ensures a relatively simple design because it avoids the formulation and solving of linear matrix inequalities.

The Takagi-Sugeno PI-FCs are designed, as in the Mamdani case, starting with the linear PI controllers to ensure the improvement in the linear control system performance. The Takagi-Sugeno FC structure is presented in Fig. 1.9A, and the input membership functions of a simple Takagi-Sugeno PI-FC are presented in Fig. 1.10.

More membership functions can be defined (e.g., those considered in Fig. 1.9B), but they complicate the rule base. One solution to deal with such situations while focusing on the design of simple fuzzy controllers is represented by fuzzy rule interpolation (Baranyi et al., 1995; Kóczy and Hirota, 1997; Yam et al., 2006; Johanyák, 2010).

The TISO-FC block presented in Fig. 1.9A is characterized, in the case of Takagi-Sugeno PI-FCs, by the weighted average method in the defuzzification module, and by the SUM and PROD operators in the inference engine. The rule base of the TISO-FC block is formulated as the decision table presented in Table 1.3, and the consequents of the fuzzy control rules are modeled by means of two functions:

$$f_{C1}(t_d) = K_P[\Delta e(t_d) + \chi \cdot e(t_d)], \quad f_{C2}(t_d) = \eta \cdot f_{C1}(t_d). \qquad (1.73a)$$

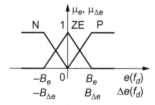

Fig. 1.10 Input membership functions of Takagi-Sugeno PI-FC. *(From David, R.-C., Precup, R.-E., Petriu, E.M., Radac, M.-B., Preitl, S., 2013. Gravitational search algorithm-based design of fuzzy control systems with a reduced parametric sensitivity. Inf. Sci. 247, 154–173.)*

Table 1.3 Rule base expressed as decision table for Takagi-Sugeno PI-FC with output integration

	$e(t_d)$		
$\Delta e(t_d)$	N	ZE	P
P	$\Delta u(t_d) = f_{C1}(t_d)$	$\Delta u(t_d) = f_{C1}(t_d)$	$\Delta u(t_d) = f_{C2}(t_d)$
ZE	$\Delta u(t_d) = f_{C1}(t_d)$	$\Delta u(t_d) = f_{C1}(t_d)$	$\Delta u(t_d) = f_{C1}(t_d)$
N	$\Delta u(t_d) = f_{C2}(t_d)$	$\Delta u(t_d) = f_{C1}(t_d)$	$\Delta u(t_d) = f_{C1}(t_d)$

The parameter η is inserted in Eq. (1.73a) to reduce the overshoot of the FCS when $e(t_d)$ and $\Delta e(t_d)$ have the same signs. Fig. 1.10 and Table 1.3 point out the tuning parameters of this simple Takagi-Sugeno PI-FC: B_e, $B_{\Delta e}$, and η.

The rule base presented in Table 1.3 can be formulated to contain only two rules because the tuning of simple Takago-Sugeno PI-FCs is generally targeted. The simplicity is ensured by the reduced number of input membership functions shown in Fig. 1.10, the symmetry of the rule base, and the simple design method dedicated to Takagi-Sugeno PI-FCs.

The modal equivalence principle leads to the following tuning equation, which is the first part of (1.71) and reduces the number of tuning parameters of the Takagi-Sugeno PI-FC to two:

$$B_{\Delta e} = \chi \cdot B_e. \tag{1.73b}$$

The design approach consists of choosing values of the parameters B_e and η according to the experience of the control system designer, and then computing $B_{\Delta e} = \alpha \cdot B_e$, $B_{\Delta u} = K_I \cdot B_e$ in terms of $B_{\Delta u}$. The domain $0 < \eta < 1$ is recommended.

1.5 Control system models and definitions of optimization problems

The FCS structure considered in this book is presented in Fig. 1.11 as a set-point filter FCS structure also given by Precup et al. (2009), where FC is the fuzzy controller, P is the process, F is the set-point filter, r is the reference input (the set-point), r_1 is the filtered reference input, d is the disturbance input, y is the controlled output, u is the control signal, and e is the control error:

$$e = r_1 - y, \tag{1.74}$$

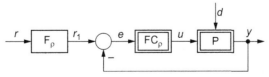

Fig. 1.11 Structure of set-point filter fuzzy control system. *(From David, R.-C., Precup, R.-E., Petriu, E.M., Radac, M.-B., Preitl, S., 2013. Gravitational search algorithm-based design of fuzzy control systems with a reduced parametric sensitivity. Inf. Sci. 247, 154–173.)*

$\boldsymbol{\rho} = \begin{bmatrix} \rho_1 & \rho_2 & \cdots & \rho_q \end{bmatrix}^T \in \mathbf{R}^q$ is the controller parameter vector with the elements ρ_γ, $\gamma = 1, \ldots, q$, which are the tuning parameters of the controller, the filter parameters can be included in this vector as indicated in Fig. 1.11, and the superscript T indicates matrix transposition. The FCS structure presented in Fig. 1.11 belongs to the two-degrees-of-freedom (2-DOF) control system structures.

As shown by Araki and Taguchi (2003), Visioli (2004), Hu et al. (2011), Vrancic and Strmcnik (2011), Iwasaki et al. (2012) Pellegrini et al. (2012), Szabo et al. (2012), Vilanova et al. (2012), Schaum and Meurer (2015), and Pachauri et al. (2017), the 2-DOF control system structures (with PI and PID controllers) have an advantage over the 1-DOF control system structures by high performance in reference input tracking and regulation in the presence of disturbance inputs. The main drawback of 2-DOF linear controllers is that the overshoot reduction is paid by a slower response with respect to the reference input. Several ways to introduce this additional block in FCS are presented by Precup et al. (2009) and Precup and Hellendoorn (2011). Inserting fuzzy logic in 2-DOF control system structures leads to improved control system performance indices in both tracking and regulation; very good results in different applications are reported by Precup et al. (2009, 2013b), Liu et al. (2012), Silveira et al. (2012), Stinean et al. (2012), and Zdešar et al. (2014).

A crisp set-point filter F will be considered in this book. The set-point filter can be considered as a fuzzy logic block as well, and accordingly designed and tuned.

The following discrete-time objective functions are defined to optimally tune the parameters of FC:

$$J_1(\boldsymbol{\rho}) = \sum_{t_d=0}^{\infty} e^2(t_d, \boldsymbol{\rho}), \tag{1.75}$$

$$J_2(\boldsymbol{\rho}) = \sum_{t_d=0}^{\infty} |e(t_d, \boldsymbol{\rho})|, \tag{1.76}$$

$$J_3(\boldsymbol{\rho}) = \sum_{t_d=0}^{\infty} \left[t_d e^2(t_d, \boldsymbol{\rho}) \right], \tag{1.77}$$

$$J_4(\boldsymbol{\rho}) = \sum_{t_d=0}^{\infty} \left[t_d | e(t_d, \boldsymbol{\rho})| \right], \tag{1.78}$$

where $J_1(\boldsymbol{\rho})$ is the sum of squared control errors, the objective function $J_2(\boldsymbol{\rho})$ is the sum of absolute control errors, $J_3(\boldsymbol{\rho})$ is the sum of squared control errors multiplied by time, and $J_4(\boldsymbol{\rho})$ is the sum of absoluter control errors multiplied by time. The vector variable of the objective functions $\boldsymbol{\rho}$ will be omitted in the sequel in certain situations for the sake of simplicity.

The convergence of the objective functions defined in Eqs. (1.75)–(1.78) requires that the steady-state values of the functions on the right-hand terms be zero. The zero steady-state value of the control error e for several types of disturbance inputs is guaranteed by controllers with an integral component.

In practical control problem solutions the sums in Eqs. (1.75)–(1.78) should be truncated to capture all transients of the FCS during the time horizon. The time horizon should include the moments when the objective functions reach their steady-state values. The upper limit of the sum depends on the dynamics of the particular process under consideration.

The minimization of the objective functions defined in Eqs. (1.75)–(1.78) aims the optimal tuning of fuzzy controllers, and is expressed in terms of the optimization problems

$$\boldsymbol{\rho}^* = \arg\min_{\boldsymbol{\rho} \in D_{\boldsymbol{\rho}}} J_1(\boldsymbol{\rho}), \tag{1.79}$$

$$\boldsymbol{\rho}^* = \arg\min_{\boldsymbol{\rho} \in D_{\boldsymbol{\rho}}} J_2(\boldsymbol{\rho}), \tag{1.80}$$

$$\boldsymbol{\rho}^* = \arg\min_{\boldsymbol{\rho} \in D_{\boldsymbol{\rho}}} J_3(\boldsymbol{\rho}), \tag{1.81}$$

$$\boldsymbol{\rho}^* = \arg\min_{\boldsymbol{\rho} \in D_{\boldsymbol{\rho}}} J_4(\boldsymbol{\rho}), \tag{1.82}$$

where $\boldsymbol{\rho}^*$ is the optimal controller parameter vector, that is, the optimal value of the vector $\boldsymbol{\rho}$, and $D_{\boldsymbol{\rho}}$ is the feasible domain of $\boldsymbol{\rho}$. Several constraints including the stability of the FCS can be imposed and expressed by means of $D_{\boldsymbol{\rho}}$. Such constraints can be expressed as several stability conditions that are derived for FCS with Mamdani fuzzy controllers (Precup and Preitl, 1999; Sugeno, 1999; Precup and Preitl, 2006; Liu et al., 2010) or with Takagi-Sugeno fuzzy controllers (Škrjanc and Blažič, 2005; Feng, 2006; Precup et al., 2007; Chang et al., 2015; Wang and Lam, 2018).

Let the process as part of servo systems be characterized by the following nonlinear continuous-time time-invariant SISO state-space model, which defines a rather general class of servo systems:

$$m(t) = \begin{cases} -1, & \text{if } u(t) \leq -u_b, \\ \dfrac{u(t) + u_c}{u_b - u_c}, & \text{if } -u_b < u(t) < -u_c, \\ 0, & \text{if } -u_c \leq |u(t)| \leq u_a, \\ \dfrac{u(t) - u_a}{u_b - u_a}, & \text{if } u_a < u(t) < u_b, \\ 1, & \text{if } u(t) \geq u_b, \end{cases} \tag{1.83}$$

$$\dot{\mathbf{x}}_P(t) = \begin{bmatrix} 0 & 1 \\ 0 & -\dfrac{1}{T_\Sigma} \end{bmatrix} \mathbf{x}_P(t) + \begin{bmatrix} 0 \\ \dfrac{k_P}{T_\Sigma} \end{bmatrix} m(t) + \begin{bmatrix} 1 \\ 0 \end{bmatrix} d(t),$$

$$y(t) = \begin{bmatrix} 1 & 0 \end{bmatrix} \mathbf{x}_P(t),$$

where t is the continuous-time argument, $t \in \mathbf{R}$, $t \geq 0$, k_P is the process gain, T_Σ is the small time constant, the control signal u is a pulse width modulated duty cycle, and m is the output of the saturation and dead zone static nonlinearity specific to the actuator. The nonlinearity is modeled by the first equation in Eq. (1.83), with the parameters u_a, u_b, and u_c, with $0 < u_a < u_b$, $0 < u_c < u_b$. The state-space model (1.83) includes the actuator and measuring element dynamics. The state vector $\mathbf{x}_P(t)$ is expressed as follows in (angular) position applications:

$$\mathbf{x}_P(t) = [x_{P,1}(t) \ \ x_{P,2}(t)]^T = [\alpha(t) \ \ \omega(t)]^T, \tag{1.84}$$

where $\alpha(t)$ is the angular position and $\omega(t)$ is the angular speed. The process structure is illustrated in Fig. 1.12.

The nonlinearity in Eq. (1.83) is neglected in the following simplified model of the process expressed as the t.f. $P(s)$:

$$P(s) = \frac{k_{EP}}{s(1 + T_\Sigma \cdot s)}. \tag{1.85}$$

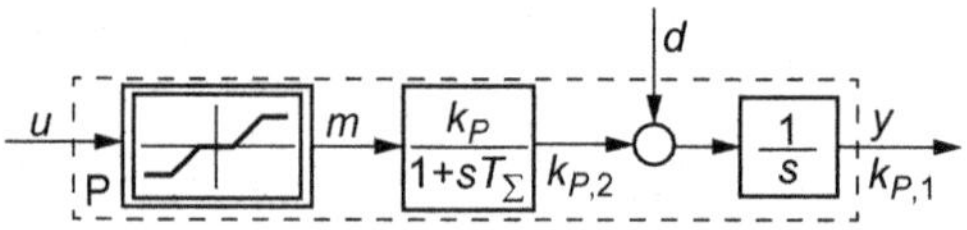

Fig. 1.12 Structure of process with saturation and dead zone static nonlinearity. *(From David, R.-C., Precup, R.-E., Petriu, E.M., Radac, M.-B., Preitl, S., 2013. Gravitational search algorithm-based design of fuzzy control systems with a reduced parametric sensitivity. Inf. Sci. 247, 154–173.)*

This t.f. is considered for u as input, y as output, and zero initial conditions. The equivalent process gain is k_{EP}.

$$k_{EP} = \begin{cases} \dfrac{k_P}{u_b - u_c}, & \text{if } -u_b < u(t) < -u_c, \\ \dfrac{k_P}{u_b - u_a}, & \text{if } u_a < u(t) < u_b. \end{cases} \quad (1.86)$$

Therefore, $P(s)$ can be used in the controller design and tuning in two cases out of the five cases concerning the nonlinearity in Eq. (1.83).

The process models given in Eqs. (1.83) and (1.85) can be employed in the control designs of servo systems in various applications accepting that the parameters k_P and T_Σ depend on the operating point. Therefore, the design of control systems with a reduced parametric sensitivity with respect to k_P and T_Σ is justified requiring other objective functions with sensitivity models included. Several approaches to the optimal tuning of fuzzy controllers with a reduced parametric sensitivity are given by David et al. (2013), and Precup et al. (2012, 2013a, b, 2014, 2017a, b) with focus on this class of servo systems.

As shown in Åström and Hägglund (1995), Preitl and Precup (1999), and Precup et al. (2009), PI controllers can cope with the process modeled by Eq. (1.85) if they are inserted in 2-DOF linear control system structures as that shown in Fig. 1.11 with a PI controller instead of FC. The transfer function of the PI controller is given in Eq. (1.65). The PI controllers can be tuned by the extended symmetrical optimum (ESO) method (Preitl and Precup, 1999) to guarantee a desired compromise to the performance specifications (i.e., maximum values of control system performance indices) imposed to the control system using a single design parameter referred to as β, with the recommended values within $1 < \beta \le 20$. The diagrams presented in Fig. 1.13 can be used in setting the value of the design parameter β and, therefore, the compromise to the control system performance indices expressed as percent overshoot σ_1 (%), settling time t_s and rise time t_r.

The PI tuning conditions specific to the ESO method are

$$k_C = \frac{1}{\sqrt{\beta} \cdot k_{EP} \cdot T_\Sigma}, \quad T_i = \beta \cdot T_\Sigma. \quad (1.87)$$

Fig. 1.13 is important because both possible values of k_{EP} according to (1.86) should be used in setting certain values of β, which ensure the fulfillment of the performance specifications imposed to the control system. A simple version of set-point filter which ensures the performance improvement of the

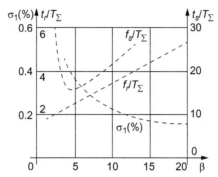

Fig. 1.13 Control system performance indices with respect to reference input vs design parameter β in the ESO method. *(From Preitl, S., Precup, R.-E., 1999. An extension of tuning relations after symmetrical optimum method for PI and PID controllers. Automatica 35 (10), 1731–1736.)*

linear control system by the cancellation of a zero in the closed-loop t.f. with respect to the reference input is characterized by the t.f.

$$F(s) = \frac{1}{1 + \beta \cdot T_\Sigma \cdot s}. \tag{1.88}$$

As shown in Section 1.4, the Takagi-Sugeno PI-FCs start with the linear PI controllers to ensure the improvement in the control system performance indices for the nonlinear process modeled in Eq. (1.83). The structure and the input membership functions of a simple Takag-Sugeno PI-FC are presented in Figs. 1.9A and 1.10, respectively.

Fig. 1.10 and Table 1.3 point out the tuning parameters of these simple Takagi-Sugeno PI-FCs: β (for the linear part of the controllers design), and B_e, $B_{\Delta e}$, and η (for the fuzzy part of the controllers design).

The application of ESO method and modal equivalence principle yields only three tuning parameters for the Takagi-Sugeno PI-FCs, $q = 3$. These parameters are included in the controller parameter vector $\boldsymbol{\rho}$ involved in the optimization problems defined in Eqs. (1.79)–(1.82):

$$\boldsymbol{\rho} = [\rho_1 \ \rho_2 \ \rho_3]^T, \quad \rho_1 = \beta, \quad \rho_2 = B_e, \quad \rho_3 = \eta. \tag{1.89}$$

The design approach dedicated to the simple Takagi-Sugeno PI-FCs with the previously defined structure consists of the following steps that result in the optimal controller parameter vector obtained by nature-inspired algorithms:

Step 1. ESO method is applied to tune the parameters of continuous-time linear PI controllers. The sampling period is set, and Tustin's method leads to Eqs. (1.64) and (1.66).

Step 2. The parameter t_{df} is set to replace ∞ in Eqs. (1.75)–(1.78) such that the finite time horizon includes all transients of the FCS until the objective functions reach the steady-state values. The feasible domains D_{ρ} are set to include all constraints imposed to the elements of $\boldsymbol{\rho}$.

Step 3. The nature-inspired algorithms are mapped onto the optimization problems (1.79)–(1.82).

Step 4. The nature-inspired algorithms are applied to compute the optimal parameter vector $\boldsymbol{\rho}^*$ and the optimal parameters as the elements of this vector

$$\boldsymbol{\rho}^* = [\rho_1^* \ \rho_2^* \ \rho_3^*]^T, \ \beta^* = \rho_1^*, \ B_e^* = \rho_2^*, \ \eta^* = \rho_3^*, \quad (1.90)$$

Next the tuning condition obtained from Eq. (1.73a) using Eqs. (1.66) and (1.86) for the optimal controller parameters leads to the optimal value of $B_{\Delta e}$, namely $B_{\Delta e}^*$:

$$B_{\Delta e}^* = \frac{2 T_s}{2 \beta^* \cdot T_{\Sigma} - T_s} B_e^*. \quad (1.91)$$

Several details concerning the application of this design approach will be presented. These details also represent the preparation for the implementation of the nature-inspired algorithms involved in steps 3 and 4, which will be described in Chapters 2–4.

The dynamic regimes considered for solving the optimization problems (1.79)–(1.82) by nature-inspired algorithms are characterized by the step-type modification of magnitude r_0 of the angular position reference input, by zero disturbance input and these regimes employ the initial state vector of the process set to the origin, $\mathbf{x}_P(0) = [0 \ 0]^T \in \mathbf{R}^2$. Other dynamic regimes characterized by different modifications of the reference input and/or of the disturbance input yield similar results but different controller tuning parameters.

The design approach and the nature-inspired algorithms are applied in this book to the design of Takagi-Sugeno PI-FCs in a case study that deals with the angular position of the experimental setup built around a direct current servo system laboratory equipment (Inteco, 2007). The experimental setup is illustrated in Figs. 1.14 and 1.15.

An optical encoder is used for the measurement of the angle and a tachogenerator for the measurement of the angular speed. The speed can also be estimated from the angle measurements. The pulse width modulation (PWM) signals that are proportional with the control signal are produced by the actuator in the power interface. The main features of the experimental setup are (Inteco, 2007): rated amplitude of 24 V, rated current of 3.1 A, rated torque of 15 N m, rated speed of 3000 rpm, and weight of inertial load

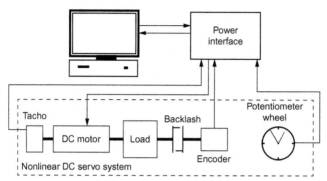

Fig. 1.14 Structure of experimental setup. *(From David, R.-C., Precup, R.-E., Petriu, E.M., Radac, M.-B., Preitl, S., 2013. Gravitational search algorithm-based design of fuzzy control systems with a reduced parametric sensitivity. Inf. Sci. 247, 154–173.)*

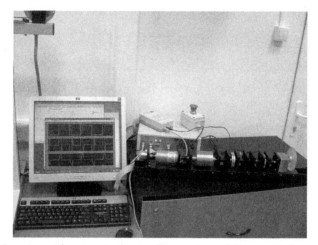

Fig. 1.15 Experimental setup in the Intelligent Control Systems Laboratory of the Politehnica University of Timisoara. *(From Precup, R.-E., Angelov, P., Costa, B.S.J., Sayed-Mouchaweh, M., 2015. An overview onfault diagnosis and nature-inspired optimal control of industrial process applications.Comput. Ind. 74, 75–94.)*

of 2.03 kg. The nominal values of the parameters of the process model given in Eqs. (1.83) and (1.85), obtained by a least-squares algorithm, are $u_a = 0.15$, $u_b = 1$, $u_c = 0.15$, $k_P = k_{EP} = 140$, and $T_\Sigma = 0.92$ s.

The variables of the objective function were initialized taking into consideration the following boundaries that define the search domain D_ρ, which is also the feasible domain of ρ:

$$D_\rho = \{\beta|\ 3 \leq \beta \leq 17\} \times \{B_e|\ 20 \leq B_e \leq 40\} \times \{\eta|\ 0.25 \leq \eta \leq 0.75\}.$$

$$(1.92)$$

The nature-inspired optimization algorithms presented in the following chapters were run for the dynamic regimes characterized by the $r = r_0 = 40$ rad step-type modification of the reference input and zero disturbance input, $d = 0$.

In order to guarantee the convergence of the objective functions for every potential solution belonging to the search domain D_ρ with the obtained fuzzy controller tuning parameters, an additional inequality-type constraint is enforced:

$$|y(t_{df}) - r(t_{df})| \le \varepsilon_y |r(t_{df}) - r(t_0)|, \qquad (1.93)$$

where t_0 is the initial time moment, t_{df} is the final time moment, and $\varepsilon_y = 0.001$ for a 2% settling time. The condition (1.93) is checked in steady-state regimes, so theoretically $t_{df} \to \infty$ as shown in Eqs. (1.75)–(1.78), but t_{df} takes practically a finite value to capture the transients in the FCS' response. As shown by Precup et al. (2017a), the condition (1.93) does not guarantee the stability of the FCS as different initial conditions can be considered.

References

Angelov, P., Yager, R., 2012. A new type of simplified fuzzy rule-based systems. Int. J. Gen. Syst. 41 (2), 163–185.

Araki, M., Taguchi, H., 2003. Two-degree-of-freedom PID controllers. Int. J. Control. Autom. Syst. 1 (4), 401–411.

Åström, K.J., Hägglund, T., 1995. PID Controllers Theory: Design and Tuning. Instrument Society of America, Research Triangle Park, NC.

Babuška, R., Verbruggen, H.B., 1996. An overview on fuzzy modeling for control. Control. Eng. Pract. 4 (11), 1593–1606.

Baranyi, P., Gedeon, T.D., Kóczy, L.T., 1995. A general method for fuzzy rule interpolation: specialized for crisp, triangular, and trapezoidal rules. In: Proceedings of 3rd European Congress on Fuzzy and Intelligent Techniques, Aachen, Germany. pp. 99–103.

Baranyi, P., Tikk, D., Yam, Y., Patton, R.J., 2003. From differential equations to PDC controller design via numerical transformation. Comput. Ind. 51 (3), 281–297.

Blažič, S., Škrjanc, I., Matko, D., 2014. A robust fuzzy adaptive law for evolving control systems. Evol. Syst. 5 (1), 3–10.

Bouallègue, S., Toumi, F., Haggège, J., Siarry, P., 2015. Advanced metaheuristics-based approach for fuzzy control systems tuning. In: Zhu, Q., Azar, A.T. (Eds.), Complex System Modelling and Control through Intelligent Soft Computations, Studies in Fuzziness and Soft Computing. In: Vol. 319. Springer International Pub-Lishing, Cham, Heidelberg, New York, Dordrecht, London, pp. 627–653.

Bühler, H., 1994. Réglage par logique floue. Presses Polytechniques et Universitaires Romandes, Lausanne.

Castillo, O., Melin, P., 2012. A review on the design and optimization of interval type-2 fuzzy controllers. Appl. Soft Comput. 12 (4), 1267–1278.

Castillo, O., Melin, P., 2014. A review on interval type-2 fuzzy logic applications in intelligent control. Inf. Sci. 279, 615–631.

Castillo, O., Valdez, F., Melin, P., 2007. Hierarchical genetic algorithms for topology optimization in fuzzy control systems. Int. J. Gen. Syst. 36 (5), 575–591.

Chang, W.-J., Kuo, C.-P., Ku, C.-C., 2015. Intelligent fuzzy control with imperfect premise matching concept for complex nonlinear multiplicative noised systems. Neurocomputing 154, 276–283.

Chang, Y.-H., Tao, C.-W., Lin, H.-W., Taur, J.-S., 2012. Fuzzy sliding-mode control for ball and beam system with fuzzy ant colony optimization. Expert Syst. Appl. 39 (3), 3624–3633.

David, R.-C., Precup, R.-E., Petriu, E.M., Radac, M.-B., Preitl, S., 2013. Gravitational search algorithm-based design of fuzzy control systems with a reduced parametric sensitivity. Inf. Sci. 247, 154–173.

Driankov, D., 2001. A reminder on fuzzy logic. In: Driankov, D., Saffiotti, A. (Eds.), Fuzzy Logic Techniques for Autonomous Vehicle Navigation. Springer-Physica Verlag, Berlin, Heidelberg, pp. 25–47.

Driankov, D., Hellendoorn, H., Reinfrank, M., 1993. An Introduction to Fuzzy Control. Springer-Verlag, Berlin, Heidelberg, New York.

Feng, G., 2006. A survey on analysis and design of model-based fuzzy control systems. IEEE Trans. Fuzzy Syst. 14 (5), 676–697.

Galichet, S., Foulloy, L., 1995. Fuzzy controllers: synthesis and equivalences. IEEE Trans. Fuzzy Syst. 3 (3), 140–148.

Gil, P., Sebastião, A., Lucena, C., 2018. Constrained nonlinear-based optimisation applied to fuzzy PID controllers tuning. Asian J. Control 20 (1), 135–148.

Guechi, E.H., Lauber, J., Dambrine, M., Klančar, G., Blažič, S., 2010. PDC control design for non-holonomic wheeled mobile robots with delayed outputs. J. Intell. Robot. Syst. 60 (3–4), 395–414.

Guerra, T.M., Sala, A., Tanaka, K., 2015. Fuzzy control turns 50: 10 years later. Fuzzy Sets Syst. 281, 168–182.

Haber, R.E., Alique, J.R., Alique, A., Hernández, J., Uribe-Etxebarria, R., 2003. Embedded fuzzy-control system for machining processes: results of a case study. Comput. Ind. 50 (3), 353–366.

Haidegger, T., Kovács, L., Precup, R.-E., Benyó, B., Benyó, Z., Preitl, S., 2012. Simulation and control for telerobots in space medicine. Acta Astronaut. 181 (1), 390–402.

Holmblad, L.P., Ostergaard, J.J., 1982. Control of a cement kiln by fuzzy logic. In: Gupta, M.M., Sanchez, E. (Eds.), Fuzzy Information and Decision Processes, North Holland, Amsterdam, pp. 389–399.

Hu, W., Xiao, G., Cai, W.-J., 2011. PID controller design based on two-degrees-of-freedom direct synthesis. In: Proceedings of 2011 Chinese Control and Decision Conference, Mianyang, China. pp. 629–634.

Iglesias, J.A., Angelov, P., Ledezma, A., Sanchis, A., 2012. Creating evolving user behavior profiles automatically. IEEE Trans. Knowl. Data Eng. 24 (5), 854–867.

Inteco, 2007. Modular Servo System, User's Manual. Inteco Ltd., Krakow.

Iwasaki, M., Seki, K., Maeda, Y., 2012. High-precision motion control techniques: a promising approach to improving motion performance. IEEE Ind. Electron. Mag. 6 (1), 32–40.

Jang, J.-S.R., Sun, C.-T., Mizutani, E., 1997. Neuro-Fuzzy and Soft Computing. A Computational Approach to Learning and Machine Intelligence, Prentice Hall, Upper Saddle River, NJ.

Johanyák, Z.C., 2010. Student evaluation based on fuzzy rule interpolation. Int. J. Artif. Intell. 5 (A10), 37–55.

Johanyák, Z.C., 2017. A modified particle swarm optimization algorithm for the optimization of a fuzzy classification subsystem in a series hybrid electric vehicle. Tech. Vjes.— Technical Gazette 24 (2), 295–301.

Kaynak, O., Erbatur, K., Ertugrul, M., 2001. The fusion of computationally intelligent methodologies and sliding-mode control—a survey. IEEE Trans. Ind. Electron. 48 (1), 4–17.

Klement, E.P., Mesiar, R., Pap, E., 2000. Triangular Norms. Kluwer Academic Publishers, Dordrecht.

Kóczy, L.T., 1996. Fuzzy if-then rule models and their transformation into one another. IEEE Trans. Syst. Man Cybern Part A 26 (5), 621–637.

Kóczy, L.T., Hirota, K., 1997. Size reduction by interpolation in fuzzy rule bases. IEEE Trans. Syst. Man Cybern. 27 (1), 14–25.

Linda, O., Manic, M., 2011. Uncertainty-robust design of interval type-2 fuzzy logic controller for delta parallel robot. IEEE Trans. Ind. Inf. 7 (4), 661–670.

Liu, G., Han, J.-H., Wu, Y.-B., Liu, M.-J., 2010. An optimal control problem of adaptive fuzzy controllers for fuzzy control systems. In: Proceedings of 2010 International Conference on Intelligent Computation Technology and Automation, Changsha, China. vol. 1, pp. 619–622.

Liu, Y., Yang, L., Duan, H., 2012. Adaptive fuzzy and H_∞ control of robotic ma-nipulators with uncertainties. Proceedings of 10th World Congress on Intelligent Control and Automation, Beijing, China, 74–79.

Mahmoodabadi, M.J., Danesh, N., 2017. Gravitational search algorithm-based fuzzy control for a nonlinear ball and beam system. J. Control Decis. 5, 229–240.

Mamdani, E.H., 1974. Applications of fuzzy control for control of simple dynamic plant. Proc. IEEE 121 (12), 1585–1588.

Mamdani, E.H., Assilian, S., 1975. An experiment in linguistic synthesis with a fuzzy logic controller. Int. J. Man Mach. Stud. 7 (1), 1–13.

Mohammadzadeh, A., Kaynak, O., Teshnehlab, M., 2014. Two-mode indirect adaptive control approach for the synchronization of uncertain chaotic systems by the use of a hierarchical interval type-2 fuzzy neural network. IEEE Trans. Fuzzy Syst. 22 (5), 1301–1312.

Noshadi, A., Shi, J., Lee, W.S., Shi, P., Kalam, A., 2016. Optimal PID-type fuzzy logic controller for a multi-input multi-output active magnetic bearing system. Neural Comput. & Applic. 27 (7), 2031–2046.

Oh, S.K., Jang, H.J., Pedrycz, W., 2011. A comparative experimental study of type-1/type-2 fuzzy cascade controller based on genetic algorithms and particle swarm optimization. Expert Syst. Appl. 38 (9), 11217–11229.

Onieva, E., Milanés, V., Villagrá, J., Pérez, J., Godoy, J., 2012. Genetic optimization of a vehicle fuzzy decision system for intersections. Expert Syst. Appl. 39 (18), 13148–13157.

Pachauri, N., Singh, V., Rani, A., 2017. Two degree of freedom PID based inferential control of continuous bioreactor for ethanol production. ISA Trans. 68, 235–250.

Passino, K.M., Yurkovich, S., 1998. Fuzzy Control. Addison-Wesley, Menlo Park, CA.

Pellegrini, E., Pletschen, N., Spirk, S., Rainer, M., Lohmann, B., 2012. Application of a model-based two-DOF control structure for enhanced force tracking in a semi-active vehicle suspension. In: Proceedings of 2012 IEEE International Conference on Control Applications, Dubrovnik, Croatia. pp. 118–123.

Pelusi, D., Mascella, R., Tallini, R., Vazquez, L., Diaz, D., 2016. Control of drum boiler dynamics via an optimized fuzzy controller. Int. J. Simul. Syst. Sci. Technol. 17 (33), 1–7.

Penedo, F., Haber, R.E., Gajate, A., del Toro, R.M., 2012. Hybrid incremental modeling based on least squares and fuzzy K-NN for monitoring tool wear in turning processes. IEEE Trans. Ind. Inf. 8 (4), 811–818.

Precup, R.-E., Angelov, P., Costa, B.S.J., Sayed-Mouchaweh, M., 2015. An overview on fault diagnosis and nature-inspired optimal control of industrial process applications. Comput. Ind. 74, 75–94.

Precup, R.-E., David, R.-C., Petriu, E.M., 2017a. Grey wolf optimizer algorithm-based tuning of fuzzy control systems with reduced parametric sensitivity. IEEE Trans. Ind. Electron. 64 (1), 527–534.

Precup, R.-E., David, R.-C., Petriu, E.M., Preitl, S., Radac, M.-B., 2012. Fuzzy control systems with reduced parametric sensitivity based on simulated annealing. IEEE Trans. Ind. Electron. 59 (8), 3049–3061.

Precup, R.-E., David, R.-C., Petriu, E.M., Preitl, S., Radac, M.-B., 2013a. Fuzzy logic-based adaptive gravitational search algorithm for optimal tuning of fuzzy-controlled servo systems. IET Control Theory Appl. 7 (1), 99–107.

Precup, R.-E., David, R.-C., Petriu, E.M., Preitl, S., Radac, M.-B., 2014. Novel adaptive charged system search algorithm for optimal tuning of fuzzy controllers. Expert Syst. Appl. 41 (4, part 1), 1168–1175.

Precup, R.-E., David, R.-C., Petriu, E.M., Radac, M.-B., Preitl, S., Fodor, J., 2013b. Evolutionary optimization-based tuning of low-cost fuzzy controllers for servo systems. Knowl.-Based Syst. 38, 74–84.

Precup, R.-E., David, R.-C., Petriu, E.M., Szedlak-Stinean, A.-I., Bojan-Dragos, C.-A., 2016. Grey wolf optimizer-based approach to the tuning of PI-fuzzy controllers with a reduced process parametric sensitivity. IFAC-PapersOnLine 48 (5), 55–60.

Precup, R.-E., David, R.-C., Preitl, S., Petriu, E.M., Tar, J.K., 2011. Optimal control systems with reduced parametric sensitivity based on particle swarm optimization and simulated annealing. In: Köppen, M., Schaefer, G., Abraham, A. (Eds.), Intelligent Computational Optimization in Engineering Techniques and Applications. Springer-Verlag, Berlin, Heidelberg, pp. 177–207.

Precup, R.-E., David, R.-C., Szedlak-Stinean, A.-I., Petriu, E.M., Dragan, F., 2017b. An easily understandable grey wolf optimizer and its application to fuzzy controller tuning. Algorithms 10 (2), 1–15. https://doi.org/10.3390/a10020068.

Precup, R.-E., Hellendoorn, H., 2011. A survey on industrial applications of fuzzy control. Comput. Ind. 62 (3), 213–226.

Precup, R.-E., Preitl, S., 1999. Fuzzy Controllers. Editura Orizonturi Universitare, Timisoara.

Precup, R.-E., Preitl, S., 2004. Optimisation criteria in development of fuzzy controllers with dynamics. Eng. Appl. Artif. Intell. 17 (6), 661–674.

Precup, R.-E., Preitl, S., 2006. Stability and sensitivity analysis of fuzzy control systems. Mechatronics applications. Acta Polytech. Hung. 3 (1), 61–76.

Precup, R.-E., Preitl, S., Petriu, E.M., Tar, J.K., Tomescu, M.L., Pozna, C., 2009. Generic two-degree-of-freedom linear and fuzzy controllers for integral processes. J. Franklin Inst. 346 (10), 980–1003.

Precup, R.-E., Tomescu, M.L., Preitl, S., Petriu, E.M., Fodor, J., Pozna, C., 2013c. Stability analysis and design of a class of MIMO fuzzy control systems. J. Intell. Fuzzy Syst. 25 (1), 145–155.

Precup, R.-E., Tomescu, M.L., Preitl, S., 2007. Lorenz system stabilization using fuzzy controllers. Int. J. Comput. Commun. Control 2 (3), 279–287.

Preitl, S., Precup, R.-E., 1997. Introducere în conducerea fuzzy a proceselor. Editura Tehnica, Bucharest.

Preitl, S., Precup, R.-E., 1999. An extension of tuning relations after symmetrical optimum method for PI and PID controllers. Automatica 35 (10), 1731–1736.

Qiu, J.-B., Gao, H.-J., Ding, S.X., 2016. Recent advances on fuzzy-model-based nonlinear networked control systems: a survey. IEEE Trans. Ind. Electron. 63 (2), 1207–1217.

Rudas, I.J., Fodor, J., 2006. Information aggregation in intelligent systems using generalized operators. Int. J. Comput. Commun. Control 1 (1), 47–57.

Rudas, I.J., Kaynak, O., 1998. New types of generalized operations. In: Kaynak, O., Turksen, B., Rudas, I.J. (Eds.), Computational Intelligence: Soft Computing and Neuro-Fuzzy Integration with Applications, NATO ASI Series F. In: vol. 192. Springer-Verlag, Berlin, Heidelberg, New York, pp. 128–156.

Sala, A., Guerra, T.M., Babuška, R., 2005. Perspectives of fuzzy systems and control. Fuzzy Sets Syst. 156 (3), 432–444.

Schaum, A., Meurer, T., 2015. Quasi-unknown input based 2-DOF control for a class of flat nonlinear SISO systems. In: Proceedings of 2015 European Control Conference, Linz, Austria, pp. 3209–3214.

Silveira, A.S., Rodríguez, J.E.N., Coelho, A.A.R., 2012. Robust design of a 2-DOF GMV controller: a direct self-tuning and fuzzy scheduling approach. ISA Trans. 51 (1), 13–21.

Stinean, A.-I., Preitl, S., Precup, R.-E., Petriu, E.M., Dragos, C.-A., Radac, M.-B., 2012. 2-DOF PI(D) Takagi-Sugeno and sliding mode controllers for BLDC drives. In: Proceedings of 15th International Power Electronics and Motion Control Conference, Novi Sad, Serbia. pp. DS2a.7-1–DS2a.7-6.

Sugeno, M., 1999. On stability of fuzzy systems expressed by fuzzy rules with singleton consequents. IEEE Trans. Fuzzy Syst. 7 (2), 201–224.

Sugeno, M., Kang, G.T., 1988. Structure identification of fuzzy model. Fuzzy Sets Syst. 28, 12–33.

Szabo, T., Buchholz, M., Dietmayer, K., 2012. Model-predictive control of powershifts of heavy-duty trucks with dual-clutch transmissions. In: Proceedings of IEEE 51st Annual Conference on Decision and Control, Maui, HI, USA, 10–13 Dec. 2012. pp. 4555–4561.

Škrjanc, I., Blažič, S., 2005. Predictive functional control based on fuzzy model: Design and stability study. J. Intell. Robot. Syst. 43 (2–4), 283–299.

Takagi, T., Sugeno, M., 1985. Fuzzy identification of systems and its application to modeling and control. IEEE Trans. Syst. Man Cybern. 15 (1), 116–132.

Teodorescu, H.-N., 2012. Taylor and bi-local piecewise approximations with neuro-fuzzy systems. Stud. Inf. Control 21 (4), 367–376.

Tsukamoto, Y., 1979. An approach to fuzzy reasoning method. In: Ragade, K., Yager, R.R. (Eds.), Advances in Fuzzy Set Theory and Applications. North-Holland, Amsterdam, pp. 137–149.

Vaščák, J., 2012. Automatic design and optimization of fuzzy inference systems. In: Zelinka, I., Snasel, V., Abraham, A. (Eds.), Handbook of Optimization: From Classical to Modern Approach. Intelligent Systems Reference Library. In: vol. 38. Springer-Verlag, Berlin, Heidelberg, pp. 287–309.

Vilanova, R., Alfaro, V.M., Arrieta, O., 2012. Simple robust autotuning rules for 2-DoF PI controllers. ISA Trans. 51 (1), 30–41.

Visioli, A., 2004. A new design for a PID plus feedforward controller. J. Process Control 14 (4), 457–463.

Vrancic, D., Strmcnik, S., 2011. Design of 2-DOF PI controller for integrating processes. In: Proceedings of 8th Asian Control Conference, Kaohsiung, Taiwan, China, pp. 1135–1140.

Vrkalovic, S., Teban, T.-A., Borlea, I.-D., 2017. Stable Takagi-Sugeno fuzzy control designed by optimization. Int. J. Artif. Intell. 15 (2), 17–29.

Wang, H.O., Tanaka, K., Griffin, M., 1995. Parallel distributed compensation of nonlinear systems by Takagi-Sugeno fuzzy model. In: Proceedings of International Joint

Conference of Fourth IEEE International Conference on Fuzzy Systems and Second International Fuzzy Engineering Symposium, Yokohama, Japan. vol. 2. pp. 531–538.

Wang, L.-K., Lam, H.-K., 2018. Local stabilization for continuous-time Takagi-Sugeno fuzzy systems with time delay. IEEE Trans. Fuzzy Syst. 26 (1), 379–385.

Yam, Y., Wong, M.L., Baranyi, P., 2006. Interpolation with function space representation of membership functions. IEEE Trans. Fuzzy Syst. 14 (3), 398–411.

Zadeh, L.A., 1965. Fuzzy sets. Inf. Control. 8 (3), 338–353.

Zdešar, A., Dovžan, D., Škrjanc, I., 2014. Self-tuning of 2 DOF control based on evolving fuzzy model. Appl. Soft Comput. 19, 403–418.

Zimmermann, H.-J., 1991. Fuzzy Set Theory—And its Applications. Kluwer Academic Publishers, Boston, Dordrecht, London.

CHAPTER 2

Nature-inspired algorithms for the optimal tuning of fuzzy controllers

Contents

2.1 Particle swarm optimization algorithms   55
2.2 Gravitational search algorithms   62
2.3 Charged system search algorithms   67
2.4 Gray Wolf optimizer algorithms   72
References   78

Abstract

This chapter describes the mechanisms occurring in four representative nature-inspired optimization algorithms: particle swarm optimization, gravitational search algorithms, charged systems search algorithms, and gray wolf optimizer algorithms. These algorithms are inserted in two steps of the design approach dedicated to the optimal tuning of simple Takagi-Sugeno proportional-integral fuzzy controllers involved in the position control of servo systems. Four optimization problems are solved and some results concerning the algorithms' behavior are outlined.

Keywords: Charged systems search algorithms, Gravitational search algorithms, Gray wolf optimizer algorithms, Particle swarm optimization, Takagi-Sugeno proportional-integral fuzzy controllers

2.1 Particle swarm optimization algorithms

Particle swarm optimization (PSO) is a population–based stochastic optimization technique that was developed and initially introduced by Kennedy and Eberhart (1995a, b). As one of the most recognized nature-inspired algorithms, PSO is inspired by the behavior of entities observed in flocks of birds or schools of fishes. The movement of the population, characterized by agents, in PSO is guided by simple laws that repeat at each iteration, helping these agents, each representing a candidate solution, flow through the multidimensional search–domain. Each age has been assigned a position

Nature-inspired Optimization Algorithms for Fuzzy Controlled Servo Systems
https://doi.org/10.1016/B978-0-12-816358-0.00002-3

vector that is updated according to the calculated velocity which takes into consideration the best position explored by the agent and best solution explored by the swarm.

As shown by Kennedy and Eberhart (1995a, b), PSO is based on two fundamental disciplines: social science and computer science. Social concepts like evaluation, comparison, and imitations of other individuals are typically associated with intelligent agents that interact in order to adapt to the environment and develop optimal patterns of behavior. Mutual learning allows individuals to become similar and transgress to more adaptive patterns of behavior. Swarm intelligence is based on the following **principles embedded in swarm intelligence optimization algorithms** (del Valle et al., 2008):

1. The proximity principle, that is, the population should be able to carry out simple time and space calculations.
2. The quality principle, that is, the population should be able to respond to quality factors in the environment.
3. The diverse response principle, that is, the population should not commit its activity to excessively long narrow channels.
4. The stability principle, that is, the population should not change its behavior every time the environment changes.
5. The adaptability principle, that is, the population should be able to change its behavior when it is worth the computational price.

The PSO algorithm starts with a random generation of candidate solutions which are continuously improved toward the optimal solutions. From this point of view PSO can be considered as an evolutionary algorithm that is similar to the genetic algorithms. The PSO algorithm uses the following computational attributes: individual particles are updated in parallel, a new value depends on the previous and its neighbors, all updates are based on the same rules. In the PSO algorithm instead of evolutionary operators, the agents are set in the D-dimensional search space search space $\mathbf{R}^D$ with randomly chosen velocities and positions knowing their best values so far and the positions in the search space $\mathbf{R}^q$. For each particle in the search space there is data about the position and velocity at each step of the iteration. The velocity of each particle is adjusted according to its previous flying experience and the experience of the other particles.

A swarm particle can be represented by the two q-dimensional vectors $\mathbf{X}_i = [x_{i1} \ x_{i2} \ \cdots \ x_{iq}]^T \in \mathbf{R}^q$ standing for the particle (also called agent) position and the particle (agent) velocity $\mathbf{V}_i = \left[v_{i1} \ v_{i2} \ \cdots \ v_{iq}\right]^T$. In addition, the best position achieved by the particle is the vector

$\mathbf{P}_{i,Best} = \begin{bmatrix} p_{i1} & p_{i2} & \cdots & p_{iq} \end{bmatrix}^T$ and the best position explored by the entire swarm so far is the vector $\mathbf{P}_{g,Best} = \begin{bmatrix} p_{g1} & p_{g2} & \cdots & p_{gq} \end{bmatrix}^T$.

As mentioned in Section 1.4, PSO algorithms are applied in steps 3 and 4 of the design approach dedicated to the simple Takagi-Sugeno Proportional-Integral fuzzy controller (PI-FCs), to solve the optimization problems defined in Eqs. (1.79)–(1.82). With this regard, **the PSO algorithms are mapped onto the optimization problems** (1.79)–(1.82) in terms of two categories of relationships:

– between the fitness function in the PSO algorithm and the objective functions (1.75)–(1.78): the fitness functions are equal to the objective functions J_Θ, $\Theta = 1, \ldots, 4$,
– between the agents' position vector $\mathbf{X}_i$ in the PSO algorithm and the parameter vector $\boldsymbol{\rho}$ of the fuzzy controller:

$$\mathbf{X}_i = \boldsymbol{\rho} \tag{2.1}$$

Since $\boldsymbol{\rho}$ is defined in accordance with Eq. (1.89), this results in a $q = 3$-dimensional search domain. The particle velocity and position update equations that govern the PSO algorithm can be expressed in terms of the state-space equations also given by Khanesar et al. (2007),

$$\mathbf{V}_i(k+1) = w(k)\mathbf{V}_i(k) + c_1 r_1\left(\mathbf{P}_{g,Best} - \mathbf{X}_i(k)\right) + c_2 r_2\left(\mathbf{P}_{i,Best} - \mathbf{X}_i(k)\right), \tag{2.2}$$

$$\mathbf{X}_i(k+1) = \mathbf{X}_i(k) + \mathbf{V}_i(k+1), \tag{2.3}$$

where r_1, r_2 are the random variables with uniform distribution between 0 and 1, i, $i = 1, \ldots, n$ are the index of the current particle in the swarm, n is the number of particles in the swarm, k, $k = 1, \ldots, k_{max}$ are the index of the current iteration, and k_{max} is the maximum number of iterations. The parameter $w(k)$ in Eq. (2.2) stands for the inertia weight, which shows the effect of the previous velocity vector on the new vector. Upper w_{max} and lower w_{min} limits are imposed to $w(k)$ in order to limit the particles movement in the search domain during the search process. The constant parameters c_1, $c_2 > 0$ represent the weighting factors of the stochastic acceleration terms that pull each particle toward their end position. Low values allow particles to roam far from the target regions before being tugged back. On the other hand, high values result in an abrupt movement toward, or past, target regions.

The individuals (particles) within the swarm learn from each other, and based on the knowledge obtained then move to become similar to their "better" previously obtained position and their "better" neighbor. The individuals within a neighborhood communicate with each other.

Different neighborhood topologies can emerge on the basis of the communication of a particle within the swarm. A star-type topology is created in the majority of cases. In that topology each particle can communicate with every other individual forming a fully connected social network, so that each particle could access the overall best position. **The PSO algorithm can be expressed according to the following steps** (Kennedy and Eberhart, 1995a, b; Khanesar et al., 2007; del Valle et al., 2008):

Step 1. Initialize the swarm placing particles at random positions inside the search domain D_ρ, set the iteration index $k=0$, set the search process iteration limit k_{max}, define the weighting factors c_1, c_2 and the inertia weight parameter $w(k)$:

$$w(k) = w_{max} - k\frac{w_{max} - w_{min}}{k_{max}} \tag{2.4}$$

The best particle position vector $\mathbf{P}_{i,Best}$ is initialized with the initial positions of the agents and the best swarm position vector $\mathbf{P}_{g,Best}$ is initialized with the position of the first agent.

Step 2. Evaluate the fitness of each particle using the objective (fitness) functions (1.75)–(1.78) based on its current position.

Step 3. Compare the performance of each individual to its best performance so far, and eventually update the best particle position vector $\mathbf{P}_{i,Best}$:

$$\mathbf{P}_{i,Best} = \mathbf{X}_i(k), \text{ if } J_\Theta(\mathbf{X}_i(k)) < J_\Theta(\mathbf{P}_{i,Best}), \ \Theta = 1, ..., 4 \tag{2.5}$$

Step 4. Compare the performance of each particle to the best global performance, and eventually update the best swarm position vector $\mathbf{P}_{g,Best}$:

$$\mathbf{P}_{g,Best} = \mathbf{X}_i(k), \text{ if } J_\Theta(\mathbf{X}_i(k)) < J_\Theta(\mathbf{P}_{g,Best}), \ \Theta = 1, ..., 4 \tag{2.6}$$

Step 5. Change the velocity of each particle according to Eq. (2.2).

Step 6. Move each particle to its new position according to Eq. (2.3).

Step 7. Increment the iteration index k and go to step 2, until the search process iteration limit k_{max} is reached.

Step 8. The algorithm is terminated, and the swarm best position $\mathbf{P}_{g,Best}$ is the final solution.

Concluding, in the context of mapping PSO algorithms onto the optimization problems (1.79)–(1.82), the solutions are the optimal parameter vector ρ^* obtained as

$$\rho^* = \mathbf{P}_{g,Best} \tag{2.7}$$

with $\mathbf{P}_{g,\ Best}$ obtained in step 8 of the PSO algorithm, and $\boldsymbol{\rho}^*$ defined in Eq. (1.90).

The simple model used in PSO has proven that it can cope with high-complexity problems as shown by Oh et al. (2011), Precup et al. (2011c, 2013), Safari et al. (2013), Pulido et al. (2014), Bouallègue et al. (2015), Johanyák (2017), and Valdez et al. (2017). In addition to the initial version of PSO, which was developed to operate with real-value search domains, an alternative version was introduced by Kennedy and Eberhart (1997) for the purpose of dealing with discrete valued search spaces. The binary version of PSO is required to deal with these finite domains.

The differences between the two versions of PSO are focused around the representation of particle position vector and movement definition. The position vector is constructed around the discrete values defined by the search domain, with the movement represented by the agent's probability of changing state in that dimension.

The flowchart of the PSO algorithm is presented in Fig. 2.1 (Precup et al., 2013).

In order to integrate the PSO-based solution in step 4 of the design approach dedicated to the simple Takagi-Sugeno PI-FCs presented in Section 1.4, the PSO algorithm parameters, mentioned in step 1 of PSO algorithm, had to be set in such a manner to achieve a prime search process. Based on the previous work presented by Precup et al. (2011c, 2013), the number of used agents $n = 20$ was set and the maximum number of iterations was set to $k_{max} = 100$. In order to have a good balance between exploration and exploitation characteristics of the algorithm the weighting parameters were set to $c_1 = 0.3$, $c_2 = 0.9$. For the setup of the inertia weight parameter $w(k)$ a linear decrease was employed throughout the interval determined by $w_{max} = 0.9$ and $w_{min} = 0.5$ according to Eq. (2.4).

Table 2.1 illustrates the values of the optimal controller tuning parameters and the minimum values of the objective functions J_1, J_2, J_3, and J_4. The dynamic regimes for the evaluation of the objective functions were specified in Section 1.4. The data presented in this table was obtained after several reruns of the algorithm that were required to deal with the arbitrary characteristic of the PSO algorithm. A more detailed analysis based on the average values of the objective functions, together with other algorithm performance indices, is presented in Chapter 5.

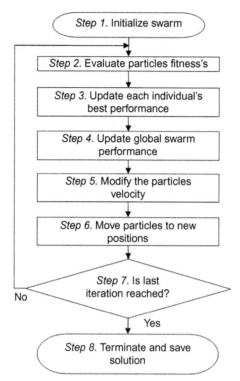

Fig. 2.1 Flowchart of particle swarm optimization algorithm. *(From Precup, R.-E., David, R.-C., Petriu, E.M., Radac, M.-B., Preitl, S., Fodor, J., 2013. Evolutionary optimization-based tuning of low-cost fuzzy controllers for servo systems. Knowl.-Based Sys. 38, 74–84.)*

Table 2.1 Results for the PSO-based minimization of J_Θ, $\Theta = 1, \dots, 4$

J_Θ	$B^*_{\Delta e}$	B^*_e	η^*	β^*	k^*_c	T^*_i	$J_{\Theta min}$
J_1	0.145191	40	0.75	3	0.004483	2.76	392,076
J_2	0.085597	40	0.75	5.08485	0.003443	4.67806	22,975.7
J_3	0.085597	40	0.75	5.08485	0.003443	4.67806	2,984,780
J_4	0.085597	40	0.75	5.08485	0.003443	4.67806	152,970

In order to have a better representation of the search process, Fig. 2.2 illustrates the evolution of parameters defined by the search domain in the case of objective function J_2.

An evolutionary display throughout the search process for all PSO algorithm's particles (agents), expressed as vector solutions ρ to the optimization problem (1.80) comprised in the search domain D_ρ, is presented in Fig. 2.3.

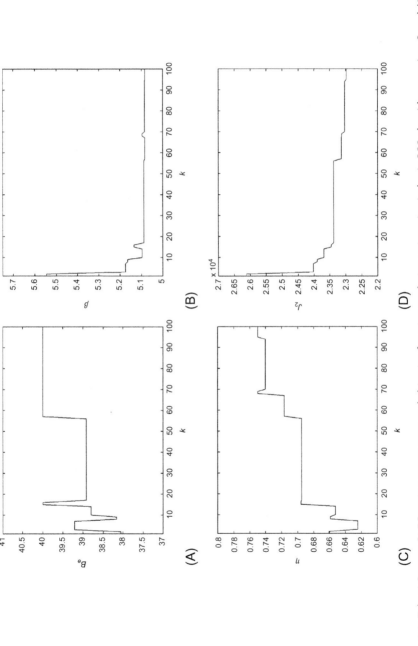

Fig. 2.2 Takagi-Sugeno PI-FC tuning parameters and objective function evolution vs iteration index in PSO algorithm running: B_e vs k (A), β vs k (B), η vs k (C), and J_2 vs k (D).

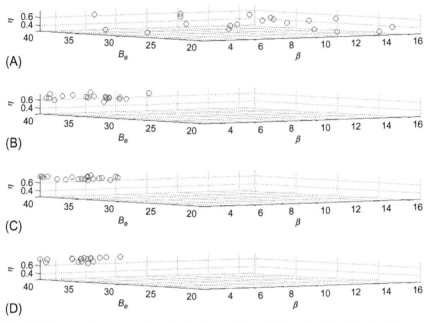

Fig. 2.3 Vector solution ρ to the optimization problem (1.80) solved by PSO algorithm in the search domain D_ρ for four values of iteration index k: $k=1$ (A), $k=15$ (B), $k=60$ (C), and $k=100$ (D).

2.2 Gravitational search algorithms

The gravitational search algorithm (GSA) (Rashedi, 2007; Rashedi et al., 2009) is inspired by Newtonian physics principals of gravity and interaction between masses. As described by the law of gravity, each agent, also referred to as object, interacts with the existing population. This interaction is proportional to each agent's mass, expressed in accordance to its fitness, and inversely proportional to the distance between them. Also, the attraction effect between the particles of the universe is introduced through the gravitational constant. The variation of the gravitational constant is modeled by the following decrease laws in relation with GSA's iterations:

$$g(k) = \psi \left(1 - \frac{k}{k_{\max}} \right) g_0, \tag{2.8}$$

$$g(k) = g_0 \exp \left(-\zeta \frac{k}{k_{\max}} \right), \tag{2.9}$$

where $g(k)$ is the value of the gravitational constant at the current iteration index k, $k_{\max}$ is the maximum number of iterations, $g_0 = g(0)$, and $\psi > 0$,

$\zeta > 0$ are parameters that are set in order to ensure GSA's convergence and to influence the search accuracy as well.

The decrease in the gravitational constant outlined in Eqs. (2.8) and (2.9) targets the modeling and the simulation of the effect of decreasing gravity. These equations show the decrease in the gravitational constant with age, which in GSA is represented by the iteration index. The GSA's convergence and search accuracy are influenced by the chosen values for parameters ψ and ζ for which the designer's experience is employed.

As mentioned by Rashedi et al. (2010), particles, referred to also as agents, are used in the GSA, and their performance is represented through their masses. The gravity force attracts each of these particles, leading to the global movement of all particles toward the particles with heavier masses. The exploitation step of the algorithm is guaranteed by the heavy masses (which correspond to good solutions, i.e., solutions close to the optimum) moving more slowly than the lighter ones.

Considering N agents and a q-dimensional search space, the position of ith agent is defined by the vector $\mathbf{X}_i$.

$$\mathbf{X}_i = \begin{bmatrix} x_i^1 & \cdots & x_i^d & \cdots & x_i^q \end{bmatrix}^T \in \mathbf{R}^q, \quad i = 1, \ldots, N, \tag{2.10}$$

where x_i^d is the position of ith agent in dth dimension, $d = 1, \ldots, q$.

The GSAs are also applied in steps 3 and 4 of the design approach dedicated to the simple Takagi-Sugeno PI-FCs given in Section 1.4 in order to solve the optimization problems defined in Eqs. (1.79)–(1.82). As in the case of PSO algorithms, **GSAs are mapped onto the optimization problems** (1.79)–(1.82) in terms of two categories of relationships:

- Between the fitness function in the GSA and the objective functions (1.75)–(1.78): the fitness functions and their corresponding fitness values are equal to the objective functions J_Θ, $\Theta = 1, \ldots, 4$.
- Between the agents' position vector $\mathbf{X}_i$ in the GSA and the parameter vector $\mathbf{\rho}$ of the fuzzy controller:

$$\mathbf{X}_i = \mathbf{\rho}. \tag{2.11}$$

The force acting on ith agent from jth agent is defined as follows at the iteration index k:

$$F_{ij}^d(k) = g(k) \frac{m_{Pi}(k) m_{Aj}(k)}{r_{ij}(k) + \varepsilon x_j^d(k)} \left[x_j^d(k) - x_i^d(k) \right], \tag{2.12}$$

where $m_{Ai}(k)$ is the active gravitational mass related to ith agent, $m_{Pj}(k)$ is the passive gravitational mass related to jth agent, $\varepsilon > 0$ is a small constant, and

$r_{ij}(k)$ is the Euclidian distance between ith and jth agents (used instead of the squared distance to simplify the GSA):

$$r_{ij}(k) = \|\mathbf{X}_i(k) - \mathbf{X}_j(k)\|. \tag{2.13}$$

To ensure the stochastic characteristic of the GSA the total force acting on ith agent in dth dimension, $F_i^d(k)$, is a randomly weighted sum of all forces exerted from the other agents:

$$F_i^d(k) = \sum_{j=1, j\neq i}^{N} \rho_j F_{ij}^d(k), \tag{2.14}$$

where ρ_j, $0 \leq \rho_j \leq 1$, is a randomly generated number. The law of motion leads to the acceleration $a_i^d(k)$ of ith agent at the iteration index k in dth dimension:

$$a_i^d(k) = \frac{F_i^d(k)}{m_{Ii}(k)}, \tag{2.15}$$

where $m_{Ii}(t)$ is the inertia mass related to ith agent.

The next velocity of an agent, $v_i^d(k+1)$, is considered as a fraction of its current velocity added to its acceleration. Therefore, the position and velocity of an agent are updated in terms of the following state-space equations (Rashedi, 2007; Rashedi et al., 2009):

$$v_i^d(k+1) = \rho_i v_i^d(k) + a_i^d(k),$$
$$x_i^d(k+1) = x_i^d(k) + v_i^d(k+1), \tag{2.16}$$

where ρ_i, $0 \leq \rho_i \leq 1$, is a uniform random variable.

The gravitational and inertial masses are (Rashedi, 2007; Rashedi et al., 2009).

$$n_i(k) = \frac{f_i(k) - w(k)}{b(k) - w(k)},$$

$$m_i(k) = \frac{n_i(k)}{\sum_{j=1}^{N} n_j(k)}, \tag{2.17}$$

$$m_{Ai} = m_{Pi} = m_{Ii} = m_i,$$

where $f_i(k)$ is the fitness value of ith agent at the iteration index k, and the terms $b(k)$ (corresponding to the best agent) and $w(k)$ (corresponding to the worst agent) are defined as follows:

$$
\begin{aligned}
b(k) &= \min_{j=1,\ldots,N} f_j(k), \\
w(k) &= \max_{j=1,\ldots,N} f_j(k).
\end{aligned}
\tag{2.18}
$$

The GSA consists of the following steps illustrated in Fig. 2.4 (Precup et al., 2013) as GSA's flowchart:

Step 1. Generate the initial population of agents, that is, initialize the q-dimensional search space, the number of agents N, set the iteration index $k=0$, set the search process iteration limit k_{max}, and initialize randomly the agents' position vector $\mathbf{X}_i(0)$.

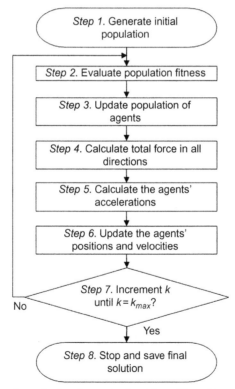

Fig. 2.4 Flowchart of gravitational search algorithm. *(From Precup, R.-E., David, R.-C., Petriu, E.M., Radac, M.-B., Preitl, S., Fodor, J., 2013. Evolutionary optimization-based tuning of low-cost fuzzy controllers for servo systems. Knowl.-Based Sys. 38, 74–84).*

Step 2. Evaluate the agents' fitness in terms of the objective (fitness) functions (1.75)–(1.78).

Step 3. Update the population of agents, that is, compute the terms $g(k)$, $b(k)$, $w(k)$, and $m_i(k)$ using Eqs. (2.8) or (2.9), (2.17), and (2.18) for $i = 1, \dots, N$.

Step 4. Calculate the total force in all directions $F_i^d(k)$, $i = 1, \dots, N$, using Eq. (2.14).

Step 5. Calculate the agents' accelerations $a_i^d(k)$ according to Eq. (2.15).

Step 6. Update the agents' velocities $v_i^d(k+1)$ and positions $x_i^d(k+1)$ using Eq. (2.16) for $i = 1, \dots, N$.

Step 7. Increment k and go to step 2 until the maximum number of iterations is reached, that is, until $k = k_{max}$.

Step 8. Stop and save the final solution in the vector $\mathbf{X}_i$ obtained so far. Employing the experience of Precup et al. (2011a, b, 2012a) and David et al. (2013) to solve the optimization problems (1.79)–(1.82), GSA's parameters were chosen in order to achieve the best search performance. The number of agents was set to $N = 20$ with a maximum number of iterations $k_{max} = 100$. Using the search domain D_ρ defined in Eq. (1.92), the dimension size was set to $q = 3$. The decrease law Eq. (2.9) of the gravitational constant was applied, with the initial value $g_0 = g(0)$ set to $g_0 = 100$ and $\zeta = 8.5$. The ε value in Eq. (2.12) was set to $\varepsilon = 10^{-4}$ in order to avoid possible divisions by zero.

The results corresponding to the four objective functions J_Θ, $\Theta = 1, \dots$, 4, are presented in Table 2.2. As in the case of other nature-inspired algorithms, the GSA solution required several reruns before obtaining the final results for each of the objective functions J_Θ, $\Theta = 1, \dots, 4$. This aspect will be approached in Chapter 5 using the average values of the objective functions together with three algorithm performance indices.

Fig. 2.5 presents a description of the evolution of the variables of the objective function and of the objective function J_2 during the search process.

In addition to Fig. 2.5, which gives a representation focused on the best position of the algorithm iterations, Fig. 2.6 exemplifies the movements of all agents used in GSA during the search process in order to better recognize the exploration and exploitation capabilities of the algorithm.

Table 2.2 Results for the GSA-based minimization of J_Θ, $\Theta = 1, \dots, 4$

J_Θ	$B_{\Delta e}^*$	B_e^*	η^*	β^*	k_c^*	T_i^*	$J_{\Theta min}$
J_1	0.138541	40	0.75	3.14374	0.004379	2.89224	390,459
J_2	0.085582	40	0.7384	5.08576	0.003443	4.6789	23,041.8
J_3	0.0845	39.5685	0.6933	5.0948	0.0034	4.6872	3,086,210
J_4	0.0848	40	0.75	5.1325	0.003427	4.72191	155,631

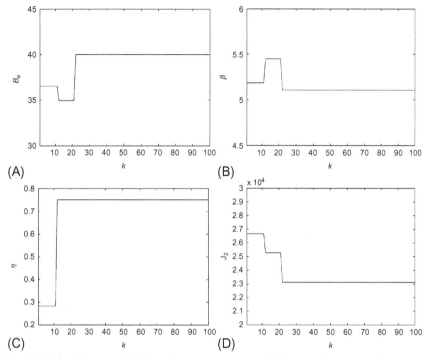

Fig. 2.5 Takagi-Sugeno PI-FC tuning parameters and objective function evolution vs iteration index in GSA running: B_e vs k (A), β vs k (B), η vs k (C), and J_2 vs k (D).

2.3 Charged system search algorithms

The specific features of charged system search (CSS) algorithms concern the random determination of the initial positions of charged particles (CPs), referred to also as agents, and the initial velocities of CPs set to zero. Each CP has an associated magnitude of charge $q_{c,i}$ and as a result it creates an electrical field around its space. The magnitude of the charge at iteration k is defined considering the quality of its solution as.

$$q_{c,i}(k) = \frac{g_i(k) - g_{best}(k)}{g_{best}(k) - g_{worst}(k)}, \quad i = 1, \dots, N, \tag{2.19}$$

where $g_{best}(k)$ and $g_{worst}(k)$ are the so far best and the worst fitness of all CPs at iteration k, $g_i(k)$ is the objective function value or the fitness function value of the ith CP at iteration k, and N is the total number of CPs. The separation distance $r_{ij}(k)$ between two CPs at iteration k is defined as (Precup et al., 2012b).

$$r_{ij}(k) = \frac{\|\mathbf{X}_i(k) - \mathbf{X}_j(k)\|}{\left\|\dfrac{(\mathbf{X}_i(k) + \mathbf{X}_j(k))}{2} - \mathbf{X}_{best}(k)\right\| + \varepsilon \mathbf{X}_i(k)}, \quad \mathbf{X}_o(k) \in \mathbf{R}^{q_s},$$

$$o \in \{i, j, best\}, \tag{2.20}$$

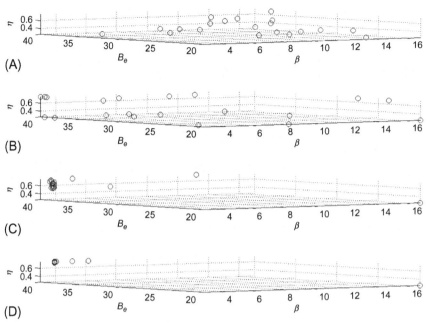

Fig. 2.6 Vector solution ρ to the optimization problem (1.80) solved by GSA algorithm in the search domain D_ρ for four values of iteration index k: $k=1$ (A), $k=15$ (B), $k=60$ (C), and $k=100$ (D).

where q_s is the dimension of the search domain, $\mathbf{X}_i(k)$ and $\mathbf{X}_j(k)$ are the position vectors of ith and jth CP at iteration k, respectively, $\mathbf{X}_{best}(k)$ is the position of the best current CP at iteration k, ε, $\varepsilon > 0$, is a constant parameter introduced to avoid singularities, and the Euclidean norm is considered in Eq. (2.20).

The electric forces between any two CPs are used for increasing the CSS algorithm's exploitation ability. The good CPs can attract other CPs and the bad ones repel the others, proportional to their rank c_{ij} (Kaveh and Talatahari, 2010a, b, c; Precup et al., 2012b) in terms of

$$c_{ij} = \begin{cases} -1, & \text{if } g_i < g_j, \\ 1, & \text{otherwise,} \end{cases} \tag{2.21}$$

where the parameter c_{ij} determines the type and the degree of influence of each CP on the other CPs, considering their fitness apart from their charges.

The value of the resultant electrical force $\mathbf{F}_i$ acting on ith CP at iteration k is (Kaveh and Talatahari, 2010a).

$$\mathbf{F}_i(k) = q_i(k) \sum_{i, i \neq j} \left(\frac{q_j(k) r_{ij}(k) i_1}{a^3} + \frac{q_j(k) i_2}{r_{ij}^2(k)} \right) c_{ij} \left(\mathbf{X}_i(k) - \mathbf{X}_j(k) \right), \quad (i_1, i_2)$$

$$= \begin{cases} (0, 1), & \text{if } r_{ij}(k) \geq a, \\ (1, 0), & \text{otherwise,} \end{cases}$$

(2.22)

where $i, j = 1, \ldots, N$. Eq. (2.22) shows that each CP is considered as a charged sphere with radius a having a uniform volume charge density.

The new position (vector) $\mathbf{X}_i(k+1)$ and velocity (vector) $\mathbf{V}_i(k+1)$ of each CP is determined in terms of Kaveh and Talatahari (2010a, b, c) and Precup et al. (2012b).

$$\mathbf{X}_i(k+1) = r_{i1} k_a(k) \left(\frac{\mathbf{F}_i}{m_i} \right) (\Delta k)^2 + r_{i2} k_v(k) \mathbf{V}_i(k) \Delta k + \mathbf{X}_i(k),$$

$$\mathbf{V}_i(k+1) = \frac{\mathbf{X}_i(k+1) - \mathbf{X}_i(k)}{\Delta k},$$

(2.23)

where k is the current iteration index which is dropped out at certain variables for the sake of simplicity, $k_a(k)$ is the acceleration parameter at iteration k, $k_v(k)$ is the velocity parameter at iteration k, which controls the influence of the previous velocity, r_{i1} and r_{i2} are two random numbers uniformly distributed in the range $0 < r_{i1}, r_{i2} < 1$, m_i is the mass of ith CP, $i = 1, \ldots, N$, which is set here as equal to $q_{c, j}$, and Δk is the time step set to 1.

The effect of the previous velocity and the resultant force acting on a CP can be decreased or increased based on the values of $k_a(k)$ and $k_v(k)$, respectively. Since $k_a(k)$ is the parameter related to the attracting forces; selecting a large value of $k_a(k)$ may cause a fast convergence and choosing a small value may increase the computational time. $k_v(k)$ controls the exploration process. The following modifications of $k_a(k)$ and $k_v(k)$ with respect to the iteration index are applied (Precup et al., 2014):

$$k_a(k) = 3 \left(1 - \frac{k}{k_{\max}} \right), \quad k_v(k) = 0.5 \left(1 + \frac{k}{k_{\max}} \right), \qquad (2.24)$$

where $k_{\max}$ is the maximum number of iterations.

The CSS algorithms are applied, as the other nature-inspired optimization algorithms treated in this book, in steps 3 and 4 of the design approach dedicated to the simple Takagi-Sugeno PI-FCs given in Section 1.4 in order to solve the optimization problems defined in Eqs. (1.79)–(1.82). As in the case of the other nature-inspired optimization algorithms, **CSS algorithms**

are mapped onto the optimization problems (1.79)–(1.82) in terms of two categories of relationships:
- Between the fitness function in the CSS algorithm and the objective functions (1.75)–(1.78): the fitness functions are equal to the objective functions J_Θ, $\Theta = 1, \ldots, 4$,
- Between the agents' position vector $\mathbf{X}_i$ in the CSS algorithm and the parameter vector $\boldsymbol{\rho}$ of the fuzzy controller:

$$\mathbf{X}_i = \boldsymbol{\rho}. \tag{2.25}$$

The CSS algorithm consists of the following steps (Precup et al., 2012b):

Step 1. Initialize the dimensional search space, the number of CPs N, set the iteration index $k = 0$, set the search process iteration limit k_{max} and randomly generate the CPs' position vector $\mathbf{X}_i(0)$, $i = 1, \ldots, N$.

Step 2. Evaluate the CPs' fitness in line with Eqs. (1.75)–(1.78).

Step 3. Update $g_{best}(k)$ and $g_{worst}(k)$, and update $q_{c,\,i}(k)$ using Eq. (2.19) for $i = 1, \ldots, N$.

Step 4. Update the values of $k_a(k)$ and $k_v(k)$ according to Eq. (2.24).

Step 5. Compute the total force in different directions using Eqs. (2.20), (2.21), and (2.22).

Step 6. Update the CPs' velocities and positions using Eq. (2.23).

Step 7. Increment k and go to step 2 until the maximum number of iterations is reached, that is, $k = k_{max}$.

Step 8. Save the optimal parameter vector as the position vector corresponding to the minimum value of objective (fitness) function.

These steps are included in the flowchart of the CSS algorithm as presented in Fig. 2.7.

The CSS algorithm previously described was employed in the optimal tuning of simple Takagi–Sugeno PI-FCs according to the design approach presented in Section 1.4. Using again the search domain D_ρ defined in Eq. (1.92), the dimension size was set to $q_s = q = 3$. Before a CSS-based solution could be used, all algorithm parameters required initialization. The number of used CPs was set to $N = 20$, the maximum number of iterations was set to $k_{max} = 100$. For the sake of simplicity each CP is considered as a charged sphere with radius $a = 1$ having a uniform volume charge density. The constant parameter ε in Eq. (2.20) was set to $\varepsilon = 10^{-4}$.

The results representing the optimal controller tuning parameters and the minimized values of the objective functions J_Θ, $\Theta = 1, \ldots, 4$, are presented in Table 2.3. A consequence of the degrees of freedom represented by the arbitrary CSS parameters requires several restarts of the search process before a final

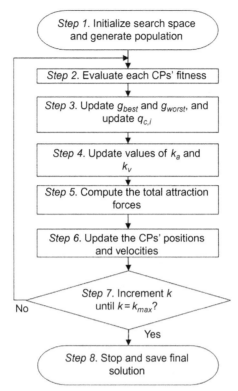

Fig. 2.7 Flowchart of charged system search algorithm. *From Precup, R.-E., David, R.-C., Petriu, E.M., Preitl, S., Radac, M.-B., 2014. Novel adaptive charged system search algorithm for optimal tuning of fuzzy controllers. Expert Sys. Appl. 41(4, part 1), 1168–1175.*

Table 2.3 Results for the CSS-based minimization of J_Θ, $\Theta = 1...4$

J_Θ	$B^*_{\Delta e}$	B^*_e	η^*	s	k^*_c	T^*_i	$J_{\Theta min}$
J_1	0.1408	40	0.75	3.0937	0.0044	2.846	390,621
J_2	0.0856	40	0.75	5.085	0.0034	4.6784	22,979.9
J_3	0.0779	36.563	0.75	5.1096	0.003435	4.701	3,089,410
J_4	0.0845	40	0.75	5.1497	0.003421	4.73768	155,252

solution can be obtained. This aspect will be analyzed in Chapter 5 based on the average values of the objective functions and other two performance indices.

Fig. 2.8 illustrates an evolutionary representation of the controller tuning parameters and of the objective function J_2 along the algorithm's iterations. Several snapshots of all CPs during the search process are shown in Fig. 2.9 in order to give a better representation of algorithm's exploration and exploitation capabilities.

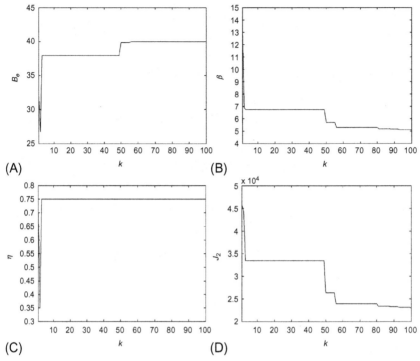

Fig. 2.8 Takagi-Sugeno PI-FC tuning parameters and objective function evolution vs iteration index in CSS algorithm running: B_e vs k (A), β vs k (B), η vs k (C), and J_2 vs k (D).

2.4 Gray wolf optimizer algorithms

The gray wolf optimizer (GWO) algorithm (Mirjalili et al., 2014) is developed on the basis of mimicking gray wolf social hierarchy and hunting habits. The social hierarchy is modeled by categorizing the population of search agents into four types of individuals, that is, alpha, beta, delta, and omega, based on their fitness. The search process is modeled by mimicking the hunting behavior of gray wolfs; it is composed of three stages in this process: searching, encircling, and attacking the prey. The first two stages are dedicated to exploration and the last one covers the exploitation. The reduced number of search parameters is an important advantage of GWO algorithms, reported in several recent applications including those reflected in various applications: benchmarks problems in optimization (Saremi et al., 2015), hyperspectral band selection (Medjahed et al., 2016), dynamic scheduling in welding industry (Lu et al., 2017), maximum power point tracking in wind energy systems (Yang et al., 2017), and optimal tuning of

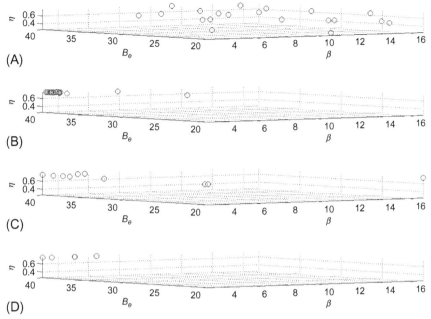

Fig. 2.9 Vector solution to the optimization problem (1.80) solved by CSS algorithm in the search domain D_ρ for four values of iteration index k: $k=1$ (A), $k=15$ (B), $k=60$ (C), and $k=100$ (D).

proportional-integral-derivative (PID)-fuzzy and PI-fuzzy controllers (Noshadi et al., 2016; Precup et al., 2016, 2017a, b).

The standard operating mechanism of GWO algorithms is given by Mirjalili et al. (2014). This process starts with the initialization of the agents (namely, gray wolves) that comprise the pack. A total number of N agents are used, and each agent has a position vector $\mathbf{X}_i(k)$ associated:

$$\mathbf{X}_i(k) = \left[x_i^1(k) \quad \cdots \quad x_i^f(k) \quad \cdots \quad x_i^q(k) \right]^T, \quad i=1,\ldots,N, \quad (2.26)$$

where $x_i^f(k)$ is the position of ith agent in the fth dimension, $f=1,\ldots,q$, k is the index of the current iteration, $k=1,\ldots,k_{max}$, and k_{max} is the maximum number of iterations.

As shown by Precup et al. (2017a), GWO algorithm's search process continues with the exploration stage represented by the search for the prey. During this stage the position of the alpha (α), beta (β), and delta (δ) agents dictate the search pattern by diverging from other agents and converging on the prey, represented by the optimal solution in the context of optimization problems solved by GWO algorithms.

The exploitation stage is represented by the attack on the prey. The top three agents constrain the other agents, viz., the omegas (ω), to update their positions according to theirs. The following notations (Precup et al., 2017a, b) are used for the top three agent position vectors, that is, the first three best solutions obtained so far (or the alpha, beta, and delta solutions):

$$\mathbf{X}^l(k) = \left[x^{l1}(k) \quad \cdots \quad x^{lf}(k) \quad \cdots \quad x^{lq}(k) \right]^T, \quad l \in \{\alpha, \beta, \delta\}. \tag{2.27}$$

The three vector solutions $\mathbf{X}^\alpha(k)$, $\mathbf{X}^\beta(k)$, and $\mathbf{X}^\delta(k)$ are obtained by the following selection process related to the objective functions J_Θ, $\Theta = 1, \ldots,$ 4, defined in Eqs. (1.75)–(1.78):

$$
\begin{aligned}
J_\Theta(\mathbf{X}^\alpha(k)) &= \min_{i=1,\ldots,N} \left\{ J_\Theta(\mathbf{X}_i(k)) | \ \mathbf{X}_i(k) \in D_\rho \right\}, \\
J_\Theta(\mathbf{X}^\beta(k)) &= \min_{i=1,\ldots,N} \left\{ J_\Theta(\mathbf{X}_i(k)) | \ \mathbf{X}_i(k) \in D_\rho \setminus \{\mathbf{X}^\alpha(k)\} \right\}, \\
J_\Theta(\mathbf{X}^\delta(k)) &= \min_{i=1,\ldots,N} \left\{ J_\Theta(\mathbf{X}_i(k)) | \ \mathbf{X}_i(k) \in D_\rho \setminus \{\mathbf{X}^\alpha(k), \mathbf{X}^\beta(k)\} \right\},
\end{aligned}
$$

$$\Theta = 1, \ldots, 4, \tag{2.28}$$

Conditions (2.28) specific to the selection process lead to the following relation between the objective functions of the top three agents:

$$J_\Theta(\mathbf{X}^\alpha(k)) < J_\Theta(\mathbf{X}^\beta(k)) < J_\Theta(\mathbf{X}^\delta(k)), \quad \Theta = 1, \ldots, 4. \tag{2.29}$$

The set of search coefficients is next defined as

$$
\begin{aligned}
a_l^f(k) &= a^f(k)\left(2r_{1l}^f - 1 \right), \\
c_l^f(k) &= 2r_{2l}^f, \quad l \in \{\alpha, \beta, \delta\},
\end{aligned}
\tag{2.30}
$$

where r_{1l}^f and r_{2l}^f are uniformly distributed random numbers that belong to the intervals $0 \le r_{1l}^f \le 1$, $0 \le r_{2l}^f \le 1$, $f = 1, \ldots, q$, and the coefficients $a^f(k)$ are linearly decreased from 2 to 0 during the search process according to

$$a^f(k) = 2[1 - (k-1)/(k_{max} - 1)], \quad f = 1, \ldots, q. \tag{2.31}$$

The approximate distances between the current solution and the alpha, beta, and delta solutions, namely $d_\alpha^{if}(k)$, $d_\beta^{if}(k)$, and $d_\delta^{if}(k)$, respectively, are defined by Precup et al. (2017b) in terms of the following relationship, which clearly points out the operations and is different to the vector operations given in other papers so far and not defined accordingly:

$$d_l^{if}(k) = |c_l^f(k) x^{lf}(k) - x_i^f(k)|, \quad i = 1, \ldots, N, \quad l \in \{\alpha, \beta, \delta\}. \tag{2.32}$$

With this regard the components (agents) $x^{lf}(k+1)$, $l \in \{\alpha, \beta, \delta\}$, of the updated alpha, beta, and delta solutions are calculated as

$$x^{lf}(k+1) = x^{lf}(k) - a_l^f(k)d_l^{if}(k), \quad f = 1, \ldots, q, \quad i = 1, \ldots, N, \quad l \in \{\alpha, \beta, \delta\},$$
$$(2.33)$$

and these components lead to the updated expressions $x_i^f(k+1)$ of agents' positions obtained as the arithmetic mean of the updated alpha, beta, and delta agents

$$x_i^f(k+1) = \left[x^{\alpha f}(k+1) + x^{\beta f}(k+1) + x^{\delta f}(k+1)\right]/3, \quad f = 1, \ldots, q,$$
$$i = 1, \ldots, N. \qquad (2.34)$$

The vector expression of Eq. (2.34) represents the following formula that gives the update on vector solution $\mathbf{X}_i(k+1)$:

$$\mathbf{X}_i(k+1) = \left[\mathbf{X}^{\alpha}(k+1) + \mathbf{X}^{\beta}(k+1) + \mathbf{X}^{\delta}(k+1)\right]/3, \quad i = 1, \ldots, N. \quad (2.35)$$

The GWO algorithm applied to solve the optimization problems (1.79)–(1.82) consists of the following steps:

Step 1. Generate the initial random gray wolf population, represented by N agents' positions in the q-dimensional search space, initialize the iteration index to $k = 0$, and set the maximum number of iterations to $k_{\max}$.

Step 2. Evaluate the performance of each member of the population of agents by simulations and/or experiments conducted on the fuzzy control system. The evaluation leads to the objective function value by mapping the GWO onto the optimization problems in terms of

$$\mathbf{X}_i(k) = \boldsymbol{\rho}, \quad i = 1, \ldots, N. \qquad (2.36)$$

Step 3. Identify the first three best solutions obtained so far, that is, $\mathbf{X}^{\alpha}(k)$, $\mathbf{X}^{\beta}(k)$, and $\mathbf{X}^{\delta}(k)$, using Eq. (2.28).

Step 4. Calculate the search coefficients using Eqs. (2.30) and (2.31).

Step 5. Move the agents to their new positions by computing $\mathbf{X}_i(k+1)$ in terms of Eqs. (2.33), (2.34), and (2.35).

Step 6. Validate the updated vector solution $\mathbf{X}_i(k+1) \in D_{\boldsymbol{\rho}}$ by checking the steady-state condition (1.93) for the fuzzy control system with the Takagi-Sugeno PI-FC parameter vector $\boldsymbol{\rho} = \mathbf{X}_i(k+1)$ obtained so far.

Step 7. Increment the iteration index k and continue with step 2 until $k_{\max}$ is reached.

Step 8. The algorithm is stopped and the final solution obtained so far is actually the solution to the optimization problems defined in Eqs. (1.79)–(1.82):

$$\boldsymbol{\rho}^* = \arg \min_{i=1,\,...,\,N} J_\Theta(\mathbf{X}_i(k_{\max})), \ \Theta = 1,\,...,\,4. \tag{2.37}$$

These eight steps are included in the flowchart of the GWO algorithm presented in Fig. 2.10.

The GWO algorithm described in this section was employed in the optimal tuning of simple Takagi-Sugeno PI-FCs according to the design approach given in Section 1.4. Using again the search domain D_ρ defined in Eq. (1.92), the dimension size was set to $q = 3$. Prior to the application of the GWO algorithm, the parameters were initialized in terms of the number of agents was set $N = 20$ and the maximum number of iterations set to $k_{\max} = 100$.

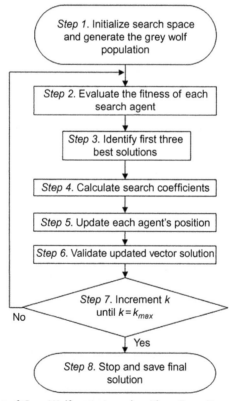

Fig. 2.10 Flowchart of Gray Wolf optimizer algorithm. *From Precup, R.-E., David, R.-C., Petriu, E.M., Szedlak-Stinean, A.-I., Bojan-Dragos, C.-A., 2016. Grey wolf optimizer-based approach to the tuning of PI-fuzzy controllers with a reduced process parametric sensitivity. IFAC-PapersOnLine 48(5), 55–60.*

Table 2.4 Results for the GWO-based minimization of J_Θ, $\Theta = 1, \ldots, 4$

J_Θ	$B^*_{\Delta e}$	B^*_e	η^*	β^*	k^*_c	T^*_i	$J_{\Theta min}$
J_1	0.1408	40	0.75	3.0937	0.0044	2.846	390,460
J_2	0.0856	40	0.75	5.085	0.0034	4.6784	22,850
J_3	0.0779	36.563	0.75	5.1096	0.003435	4.701	3,088,145
J_4	0.0845	40	0.75	5.1497	0.003421	4.73768	155,037

The results representing the optimal controller tuning parameters and the minimized values of the objective functions J_Θ, $\Theta = 1, \ldots, 4$, are presented in Table 2.4. A consequence of the degrees of freedom represented by the arbitrary CSS parameters requires several restarts of the search process before a final solution can be obtained. This aspect will be analyzed in Chapter 5 based on the average values of the objective functions and other two performance indices.

In order to have a better representation of the search process, Fig. 2.11 illustrates the evolution of parameters defined by the search domain in the

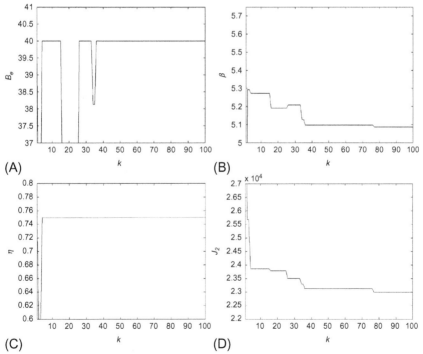

Fig. 2.11 Evolution of Takagi-Sugeno PI-FC tuning parameters and objective function evolution vs iteration index of GWO algorithm running: B_e vs k (A), β vs k (B), η vs k (C), and J_2 vs k (D).

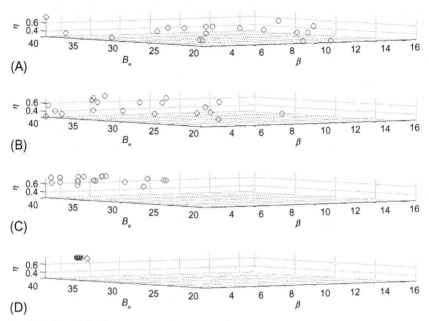

Fig. 2.12 Search for best vector solution ρ to the optimization problem (1.80) solved by GWO algorithm in the search domain D_ρ for four values of iteration index k: $k=1$ (A), $k=15$ (B), $k=60$ (C), and $k=100$ (D).

case of objective function J_2. An evolutionary display throughout the search process for all GWO algorithm's vector solutions ρ to the optimization problem Eq. (1.80) belonging to the search domain D_ρ is presented in Fig. 2.12.

References

Bouallègue, S., Toumi, F., Haggège, J., Siarry, P., 2015. Advanced metaheuristics-based approach for fuzzy control systems tuning. In: Zhu, Q., Azar, A.T. (Eds.), Complex System Modeling and Control Through Intelligent Soft Computations, Studies in Fuzziness and Soft Computing. In: 319, Springer International Publishing, Cham, pp. 627–653.

David, R.-C., Precup, R.-E., Petriu, E.M., Radac, M.-B., Preitl, S., 2013. Gravitational search algorithm-based design of fuzzy control systems with a reduced parametric sensitivity. Inform. Sci. 247, 154–173.

del Valle, Y., Venayagamoorthy, G.K., Mohagheghi, S., Hernandez, J.C., Harley, R.G., 2008. Particle swarm optimization: basic concepts, variants and applications in power systems. IEEE Trans. Evol. Comput. 12 (2), 171–195.

Kennedy, J., Eberhart, R.C., 1995a. Particle swarm optimization. In: Proceedings of 1995 IEEE International Conference on Neural Networks, Perth, Australia, pp. 1942–1948.

Johanyák, Z.C., 2017. A modified particle swarm optimization algorithm for the optimization of a fuzzy classification subsystem in a series hybrid electric vehicle. Technicki Vjesnik – Tech. Gaz. 24 (2), 295–301.

Kaveh, A., Talatahari, S., 2010a. A novel heuristic optimization method: charged system search. Acta Mech. 213, 267–289.

Kaveh, A., Talatahari, S., 2010b. Optimal design of truss structures via the charged system search algorithm. Struct. Multidiscip. Optim. 37 (6), 893–911.

Kaveh, A., Talatahari, S., 2010c. A charged system search with a fly to boundary method for discrete optimum design of truss structures. Asian J. Civ. Eng. (Building and Housing) 11 (3), 277–293.

Kennedy, J., Eberhart, R.C., 1995b. A new optimizer using particle swarm theory. In: Proceedings of 6th International Symposium on Micro Machine and Human Science, Nagoya, Japan, pp. 39–43.

Kennedy, J., Eberhart, R.C., 1997. A discrete binary version of the particle swarm algorithm. In: Proceedings of 1997 IEEE International Conference on Systems, Man, and Cybernetics, Orlando, FL, USA. vol. 5, pp. 4104–4108.

Khanesar, M.A., Tavakoli, H., Teshnehlab, M., Shoorehdeli, M.A., 2007. A novel binary particle swarm optimization. In: Proceedings of 2007 Mediterranean Conference on Control & Automation, Athens, Greece, pp. 1–6.

Lu, C., Gao, L., Li, X.-Y., Xiao, S.-Q., 2017. A hybrid multi-objective gray wolf optimizer for dynamic scheduling in a real-world welding industry. Eng. Appl. Artif. Intel. 57, 61–79.

Medjahed, S.A., Saadi, T.A., Benyettou, A., Ouali, M., 2016. Gray wolf optimizer for hyperspectral band selection. Appl. Soft Comput. 40, 178–186.

Mirjalili, S., Mirjalili, S.M., Lewis, A., 2014. Grey wolf optimizer. Adv. Eng. Softw. 69, 46–61.

Noshadi, A., Shi, J., Lee, W.S., Shi, P., Kalam, A., 2016. Optimal PID-type fuzzy logic controller for a multi-input multi-output active magnetic bearing system. Neural Comput. Applic. 27 (7), 2031–2046.

Oh, S.K., Jang, H.J., Pedrycz, W., 2011. A comparative experimental study of type-1/type-2 fuzzy cascade controller based on genetic algorithms and particle swarm optimization. Expert Syst. Appl. 38 (9), 11,217–11,229.

Precup, R.-E., David, R.-C., Petriu, E.M., 2017a. Grey wolf optimizer algorithm-based tuning of fuzzy control systems with reduced parametric sensitivity. IEEE Trans. Ind. Electron. 64 (1), 527–534.

Precup, R.-E., David, R.-C., Petriu, E.M., Preitl, S., Radac, M.-B., 2012a. Novel adaptive gravitational search algorithm for fuzzy controlled servo systems. IEEE Trans. Ind. Inf. 8 (4), 791–800.

Precup, R.-E., David, R.-C., Petriu, E.M., Preitl, S., Radac, M.-B., 2012b. Charged system search algorithms for optimal tuning of PI controllers. IFAC Proc. Vol. 45 (4), 115–120.

Precup, R.-E., David, R.-C., Petriu, E.M., Preitl, S., Radac, M.-B., 2014. Novel adaptive charged system search algorithm for optimal tuning of fuzzy controllers. Expert Sys. Appl. 41 (4, part 1), 1168–1175.

Precup, R.-E., David, R.-C., Petriu, E.M., Radac, M.-B., Preitl, S., Fodor, J., 2013. Evolutionary optimization-based tuning of low-cost fuzzy controllers for servo systems. Knowl.-Based Sys. 38, 74–84.

Precup, R.-E., David, R.-C., Petriu, E.M., Szedlak-Stinean, A.-I., Bojan-Dragos, C.-A., 2016. Grey wolf optimizer-based approach to the tuning of PI-fuzzy controllers with a reduced process parametric sensitivity. IFAC-PapersOnLine 48 (5), 55–60.

Precup, R.-E., David, R.-C., Petriu, E.M., Preitl, S., Paul, A.S., 2011a. Gravitational search algorithm-based tuning of fuzzy control systems with a reduced parametric sensitivity.

In: Gaspar-Cunha, A., Takahashi, R., Schaefer, G., Costa, L. (Eds.), Advances in Intelligent and Soft Computing. In: vol. 96. Springer-Verlag, Berlin, Heidelberg, pp. 141–150.

Precup, R.-E., David, R.-C., Petriu, E.M., Preitl, S., Radac, M.-B., 2011b. Gravitational search algorithms in fuzzy control systems tuning. IFAC Proc. Vol. 44 (1), 13624–13629.

Precup, R.-E., David, R.-C., Preitl, S., Petriu, E.M., Tar, J.K., 2011c. Optimal control systems with reduced parametric sensitivity based on particle swarm optimization and simulated annealing. In: Köppen, M., Schaefer, G., Abraham, A. (Eds.), Intelligent Computational Optimization in Engineering Techniques and Applications. Studies in Computational Intelligence, 366, Springer-Verlag, Berlin, Heidelberg, pp. 177–207.

Precup, R.-E., David, R.-C., Szedlak-Stinean, A.-I., Petriu, E.M., Dragan, F., 2017b. An easily understandable grey wolf optimizer and its application to fuzzy controller tuning. Algorithms 10 (2), 1–15. https://doi.org/10.3390/a10020068.

Pulido, M., Melin, P., Castillo, O., 2014. Particle swarm optimization of ensemble neural networks with fuzzy aggregation for time series prediction of the Mexican Stock Exchange. Inform. Sci. 280, 188–204.

Rashedi, E., 2007. Gravitational Search Algorithm. M.Sc. Thesis, Shahid Bahonar University of Kerman, Kerman, Iran.

Rashedi, E., Nezamabadi-pour, H., Saryazdi, S., 2009. GSA: a gravitational search algorithm. Inform. Sci. 179 (13), 2232–2248.

Rashedi, E., Nezamabadi-pour, H., Saryazdi, S., 2010. BGSA: binary gravitational search algorithm. Nat. Comput. 9 (3), 727–745.

Safari, S., Ardehali, M.M., Sirizi, M.J., 2013. Particle swarm optimization based fuzzy logic controller for autonomous green power energy system with hydrogen storage. Energ. Conver. Manage. 65, 41–49.

Saremi, S., Mirjalili, S.Z., Mirjalili, S.M., 2015. Evolutionary population dynamics and grey wolf optimizer. Neural Comput. Applic. 26 (5), 1257–1263.

Valdez, F., Vázquez, J.C., Melin, P., Castillo, O., 2017. Comparative study of the use of fuzzy logic in improving particle swarm optimization variants for mathematical functions using co-evolution. Appl. Soft Comput. 52, 1070–1083.

Yang, B., Zhang, X.-S., Yu, T., Shu, H.-C., Fang, Z.-H., 2017. Grouped grey wolf optimizer for maximum power point tracking of doubly-fed induction generator based wind turbine. Energ. Conver. Manage. 133, 427–443.

CHAPTER 3

Adaptive nature-inspired algorithms for the optimal tuning of fuzzy controllers

Contents

3.1 Adaptive gravitational search algorithms   81
3.2 Adaptive charged system search algorithms   86
3.3 Fuzzy logic-based adaptive gravitational search algorithms   92
References   99

Abstract

This chapter describes the adaptation mechanisms in two representative nature-inspired optimization algorithms, namely gravitational search algorithms (GSAs) and charged systems search algorithms. The adaptation in both algorithms is carried out in terms of the 5E learning cycle, and fuzzy logic is inserted in GSAs to adapt two parameters. These algorithms are inserted in the design approach dedicated to the optimal tuning of simple Takagi-Sugeno proportional-integral fuzzy controllers for the position control of servo systems. Four optimization problems are solved and representative results concerning the algorithms' behavior are outlined.

Keywords: Adaptive charged systems search algorithms, Adaptive gravitational search algorithms, Fuzzy logic-based adaptation, Popov sum, Takagi-Sugeno proportional-integral fuzzy controllers

3.1 Adaptive gravitational search algorithms

Although the standard gravitational search algorithm (GSA) shows promising results illustrated by David et al. (2013), Precup et al. (2013c, 2015), and Mahmoodabadi and Danesh (2018) with focus on the optimal tuning of fuzzy controllers and fuzzy models, along with a good computational efficiency and ease of implementation, it uses several predefined parameters and schedules which fail to take into consideration the state of the search process. Hence, the algorithm can become computationally inefficient as the exploration–exploitation ratio could not efficient correlate with the

Nature-inspired Optimization Algorithms for Fuzzy Controlled Servo Systems
https://doi.org/10.1016/B978-0-12-816358-0.00003-5

algorithm's state, thus increasing the probability of getting trapped in local minima situations.

Therefore, the most important and interesting goals in GSA development are optimal resource usage and avoiding local optima. The adaptive GSA proposed by Precup et al. (2012a, 2013a) offers a superior search process compared to the standard GSA by improving the exploration of the search space as it continues the development of stage-based adaptation of algorithm parameters (Liu et al., 2010). This improvement is ensured by the use of a learning model for the algorithm, inspired by the 5E learning cycle discussed by Bybee (2002) and Balci et al. (2006).

As the standard GSA, the adaptive version is governed by the same operating mechanism which is based on the use of agents (i.e., particles) and on Newton's law of gravity (Rashedi et al., 2009, 2010). **The algorithm consists of the following stages** (Precup et al., 2012a, 2013a, 2014b):

I. **Engagement**. The initial N agents' number is defined and their positions are generated randomly inside the bounds of the search domain D_ρ:

$$\mathbf{X}_i = \begin{bmatrix} x_i^1 & \cdots & x_i^d & \cdots & x_i^q \end{bmatrix}^T, \quad \mathbf{X}_i = \mathbf{\rho}, \quad i = 1,\ldots,N, \tag{3.1}$$

where $\mathbf{X}_i$ is agents' position vector, x_i^d is the position of ith agent in dth dimension of the $q=3$-dimensional search space, as results from Eq. (1.89) in order to solve the optimization problems Eqs. (1.79)–(1.82). The maximum number of iterations k_{max} of the search process is set and the iteration index k is set to $k=0$ and will be incremented at the end of the iteration according to step 7 in Fig. 2.4.

II. **Exploration**. This stage allows the algorithm to discover the extent of the search space. The following linear decrease law of the gravitational constant is employed:

$$g(k) = g_0 \left(1 - \psi \frac{k}{k_{max}} \right), \tag{3.2}$$

where $g(k)$ is the value of gravitational constant at current iteration index k, g_0 is the initial $g(k)$, and $\psi > 0$ is an a priori set parameter which ensures a trade-off to GSA's convergence and search accuracy.

The agent's velocities and positions are updated using the state-space-type equations

$$\begin{aligned} v_i^d(k+1) &= \rho_i v_i^d(k) + a_i^d(k), \\ x_i^d(k+1) &= x_i^d(k) + v_i^d(k+1), \end{aligned} \tag{3.3}$$

where ρ_i, $0 \leq \rho_i \leq 1$, is a uniform random variable; $a_i^d(k)$ is the acceleration of ith agent in dth dimension, defined as

$$a_i^d(k) = \frac{1}{m_{Ii}(k)} \sum_{j=1, j \neq i}^{N} \sigma_j \frac{g(k) m_{Pi}(k) m_{Aj}(k) \left[x_j^d(k) - x_i^d(k) \right]}{r_{ij}(k) + \varepsilon x_j^d(k)}, \qquad (3.4)$$

with σ_j, $0 \leq \sigma_j \leq 1$ a random generated number, $m_{Ii}(k)$, $m_{Pi}(k)$, and $m_{Aj}(k)$ are the inertial, passive, and active gravitational masses related to ith and jth agent, $\varepsilon > 0$ is a relatively small parameter, and $r_{ij}(k)$ is the Euclidian distance between ith and jth agents:

$$r_{ij}(k) = \|\mathbf{X}_i(k) - \mathbf{X}_j(k)\|. \qquad (3.5)$$

The active gravitational mass and the inertial mass are calculated in terms of

$$n_i(k) = \frac{f_i(k) - \max_{j=1,\ldots,N} f_j(k)}{\min_{j=1,\ldots,N} f_j(k) - \max_{j=1,\ldots,N} f_j(k)},$$

$$m_i(k) = \frac{n_i(k)}{\sum_{j=1}^{N} n_j(k)}, \qquad (3.6)$$

$$m_{Ai} = m_{Ii} = m_i.$$

Stage II of the adaptive GSA is carried out for the first 15% iterations (i.e., runs) in the search process.

III. **Explanation**. Algorithm's parameters restrict agents' movement during the next 45% iterations in the search process, by the introduction of a more aggressive decrease law of $g(k)$ according to

$$g(k) = g_0 \exp \left(-\zeta \frac{k}{k_{max}} \right), \qquad (3.7)$$

where $\zeta > 0$ is an a priori set parameter, which affects GSA's convergence and search accuracy.

IV. **Elaboration**. The remaining 40% of iterations are characterized by setting the general position for the optimal value of the fitness function and leaving the remaining time to refine the obtained results. The value of $g(k)$ stops decreasing, and during this stage the worst agents' positions are reset to the best values obtained so far after each run.

V. Evaluation. The tuned parameters, obtained at the end of the search process, are applied to the real-world optimization problem in order to evaluate the quality of the solution.

Stages I–V of the adaptive GSA are different than those presented in the nonadaptive GSA, which consist of the stages I, III, and V, but with a single and fixed constant decrease in law of the gravitational constant and without resetting the worst fitness and position.

The computational complexity of this optimization algorithm is influenced by the complexity of the evaluation of the objective function (viz., the fitness function) and by the number of runs of the GSA. For the objective functions considered in this book, variable (i.e., tuning parameter of the fuzzy controller) is altered separately at every stage and a new evaluation of the objective function is required in order to determine the variation impact. This results in an increased number of evaluations for the objective functions, which is proportional to the number of variables. In the case of this adaptive GSA the computational complexity varies depending on the stage of the search process that is actually involved. Stage III is the most computationally intensive one because two parameters are modified simultaneously.

Fig. 3.1 illustrates the adaptive GSA's stages.

Introducing this adaptive GSA in step 4 of the design approach dedicated to the simple Takagi-Sugeno proportional-integral fuzzy controllers (PI-FCs) presented in Section 1.4 requires setting the algorithm's parameters. The number of used agents $N=20$ was defined and the maximum number of iterations was set to $k_{max}=100$. The ψ and g_0 parameters in Eq. (3.2) were set as $\psi=0.5$ and $g_0=100$. The ε parameter in Eq. (3.4) was set to $\varepsilon=10^{-4}$ in order to avoid possible divisions by zero. The ζ parameter in Eq. (3.7) was set to $\zeta=8.5$, and g_0 in Eq. (3.7) was kept the same as in Eq. (3.2).

Table 3.1 illustrates the values of the optimal controller tuning parameters and the minimum values of the objective functions J_1, J_2, J_3, and J_4 related to their adaptive GSA-based minimization. The dynamic regimes for the evaluation of the objective functions were specified in Section 1.4. The data presented in this table was obtained after several reruns of the algorithm that were required to deal with the arbitrary characteristic of the adaptive GSA. A more detailed analysis based on the average values of the objective functions, together with other algorithm performance indices is presented in Chapter 5.

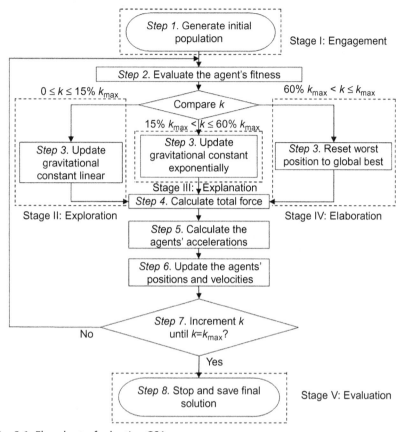

Fig. 3.1 Flowchart of adaptive GSA.

Table 3.1 Results for the adaptive GSA-based minimization of J_Θ, $\Theta = 1, \ldots, 4$

J_Θ	$B^*_{\Delta e}$	B^*_e	η^*	β^*	k^*_c	T^*_i	$J_{\Theta min}$
J_1	0.1386	40	0.75	3.143	0.004379	2.8917	390,459
J_2	0.0843	39.4214	0.75	5.0872	0.0034	4.6802	23,042.5
J_3	0.0851	40	0.75	5.1118	0.0034	4.70288	3,025,720
J_4	0.0856	39.9958	0.75	5.0849	0.0034	4.678	2,984,880

In order to have a better representation of the search process, Fig. 3.2 illustrates the evolution of parameters defined by the search domain in the case of objective function J_2.

Fig. 3.3 illustrates the evolution of all agents' positions for the adaptive GSA after the first four stages of the search process. These positions are

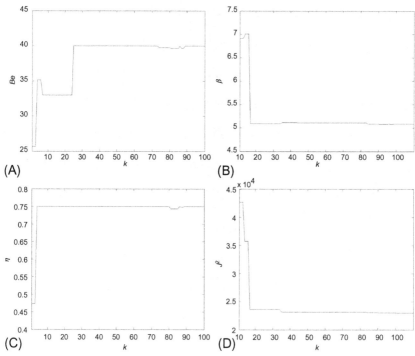

Fig. 3.2 Takagi-Sugeno PI-FC tuning parameters and objective function J_2 evolution vs iteration index in adaptive GSA running: B_e vs k (A), β vs k (B), η vs k (C), and J_2 vs k (D).

expressed as vector solutions $\boldsymbol{\rho}$ to the optimization problem (1.80), and they belong to the search domain $D_{\boldsymbol{\rho}}$.

3.2 Adaptive charged system search algorithms

As in the case of adaptive GSA, the same framework can be integrated for the classical version of the charged system search (CSS) algorithm. The adaptation uses the specific features of CSS algorithms, which are based on the interactions between charged particles (CPs) as they are moving through a predefined search domain, starting with arbitrarily determined initial positions with zero initial velocities. The CP is characterized by an associated magnitude of charge q_i and as a result it creates an electrical field around its space. The magnitude of the charge at iteration k is defined considering the quality of its solution as

$$q_{c,i}(k) = \frac{g_i(k) - g_{best}(k)}{g_{best}(k) - g_{worst}(k)}, \quad i = 1, \ldots, N, \tag{3.8}$$

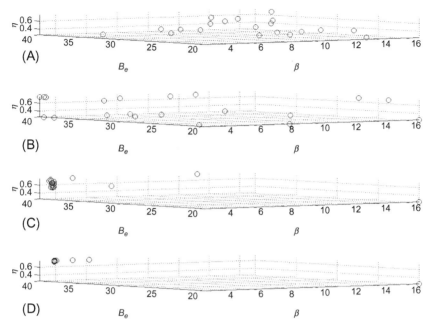

Fig. 3.3 Evolution of vector solution ρ to the optimization problem (1.80) solved by adaptive GSA in the search domain D_ρ for four values of iteration index k: $k=1$ (A), $k=15$ (B), $k=60$ (C), and $k=100$ (D).

where $g_{best}(k)$ and $g_{worst}(k)$ are the so far best and the worst fitness of all CPs at iteration k, $g_i(k)$ is the objective function value or the fitness function value of ith CP at iteration k, and N is the total number of CPs. The separation distance r_{ij} between two CPs at iteration k is defined as

$$r_{ij}(k) = \frac{\|\mathbf{X}_i(k) - \mathbf{X}_j(k)\|}{\left\|\dfrac{\mathbf{X}_i(k) + \mathbf{X}_j(k)}{2} - \mathbf{X}_{best}(k)\right\| + \varepsilon \mathbf{X}_i(k)}, \quad \mathbf{X}_o(k) \in \mathbf{R}^{q_s},$$

$$o \in \{i, j, best\}, \tag{3.9}$$

where $\mathbf{X}_i(k)$ and $\mathbf{X}_j(k)$ are the positions of ith and jth CP at iteration k, respectively, $\mathbf{X}_{best}(k)$ is the position of the best current CP at iteration k, and the relative small parameter $\varepsilon > 0$ is introduced to avoid singularities.

For the optimization problems defined in Eqs. (1.79)–(1.82), as shown in the controller parameter vector ρ given in Eq. (1.89), three tuning parameters of the Takagi-Sugeno PI-FCs are used resulting in a $q = q_s = 3$-dimensional search space.

The exploitation ability of CSS algorithms is increased by the electric forces between the CPs. When a search space is a noisy domain, having a complete search before converging to a result is necessary. In such a condition, the addition of the ability of repelling forces to the algorithm may improve its performance. Good CPs can attract other agents and bad CPs repel the others, proportional to their rank c_{ij}

$$c_{ij} = \begin{cases} -1 & \text{if } f_i < f_j, \\ 1 & \text{otherwise.} \end{cases} \tag{3.10}$$

The rank c_{ij} sets the type and the degree of influence of each CP on the other CPs considering their fitness function values apart from their charges. This means that good agents are awarded the capability of attraction and bad ones are given the repelling feature, resulting in the improvement of the exploration and exploitation abilities of the algorithm. When a good agent attracts a bad one, the exploitation ability is provided for the algorithm; on the other hand, when a bad agent repels a good CP, the exploration is provided.

The expression of the resultant electrical force F_i acting on the ith CP at iteration k is

$$\mathbf{F}_i(k) = q_i(k) \sum_{j=1, j\neq i}^{N} \left(\frac{q_j(k) r_{ij}(k) i_1}{a^3} + \frac{q_j(k) i_2}{r_{ij}^2(k)} \right) c_{ij} \left(\mathbf{X}_i(k) - \mathbf{X}_j(k) \right), \quad (i_1, i_2)$$

$$= \begin{cases} (0, 1), & \text{if } r_{ij}(k) \geq a, \\ (1, 0), & \text{otherwise.} \end{cases}$$

$$\tag{3.11}$$

Eq. (3.11) indicates that each CP is considered as a charged sphere with radius a having a uniform volume charge density.

The new position of ith CP, $\mathbf{X}_i(k+1)$, and the new velocity of ith CP, $\mathbf{V}_i(k+1)$, are obtained in terms of (Kaveh and Talatahari, 2010a, b, c; Precup et al., 2014a, b)

$$\mathbf{X}_i(k+1) = r_{i1}(k) k_a(k) \left(\frac{\mathbf{F}_i(k)}{m_i(k)} \right) (\Delta k)^2 + r_{i2} k_v(k) \mathbf{V}_i(k) \Delta k + \mathbf{X}_i(k),$$

$$\tag{3.12}$$

$$\mathbf{V}_i(k+1) = \frac{\mathbf{X}_i(k+1) - \mathbf{X}_i(k)}{\Delta k},$$

where k is the current iteration index, $k_a(k)$ is the acceleration parameter, $k_v(k)$ is the velocity parameter, r_{i1} and r_{i2} are two random numbers uniformly distributed in the range of $(0, 1)$, m_i is the mass of the ith CP, $i = 1, \dots, N$, $m_i = q_i$ in the sequel, and Δk is the time step set to 1.

The effect of previous velocity and the resultant force acting on a CP can be decreased or increased on the basis of the values of $k_v(k)$ and $k_a(k)$, respectively. Excessive search in the early iterations may improve the exploration ability; however, a gradual decrease is advised by Kaveh and Talatahari (2010a, b, c). Since $k_a(k)$ is the parameter related to the attracting forces, selecting a large value for this parameter may cause a fast convergence and choosing a small value can increase the computation time. In fact, $k_a(k)$ is a control parameter of the exploitation; therefore, choosing an incremental function can improve the performance of the algorithm. In addition, the direction of the pervious velocity of a CP is not necessarily the same as the resultant force. Thus, it can be concluded that the velocity parameter $k_v(k)$ controls the exploration process, so an increasing function can be selected. Therefore, based on extended experimental practice, we suggest the following modifications of $k_a(k)$ and $k_v(k)$ with respect to the iteration index k (Precup et al., 2014a):

$$k_a(k) = 3\left(1 - \frac{k}{k_{\max}}\right), \quad k_v(k) = 0.5\left(1 + \frac{k}{k_{\max}}\right), \tag{3.13}$$

where $k_{\max}$ is the maximum number of iterations.

The adaptive CSS algorithm is expressed in terms of five stages, I–V, illustrated in Fig. 3.4 and described as follows:

I. **Engagement**. This stage is dedicated to the initialization of adaptive CSS algorithm's population and parameters.

II. **Exploration**. The adaptive CSS algorithm is run with no modifications of $k_a(k)$ and $k_v(k)$, so no constraints are applied to CPs' movements. This stage accounts for the first 20% of $k_{\max}$ iterations.

III. **Explanation**. The adaptive CSS algorithm is run using the linear modifications of k_a and k_v according to Eq. (3.13). The next 40% $k_{\max}$ iterations are assigned to this stage. The adaptation is obtained on the basis of experience in adaptive GSA algorithm (Precup et al., 2012b).

IV. **Elaboration**. This stage uses the last 40% of $k_{\max}$ iterations in adaptive CSS algorithm's search process runs. In addition, at each run the agent's position with the worst fitness is reset to the position of the agent with the best fitness.

V. **Evaluation**. CPs' positions are mapped onto the variables of the optimization problem, and the objective functions are evaluated using the real-world model of the optimization problem to evaluate the obtained solution.

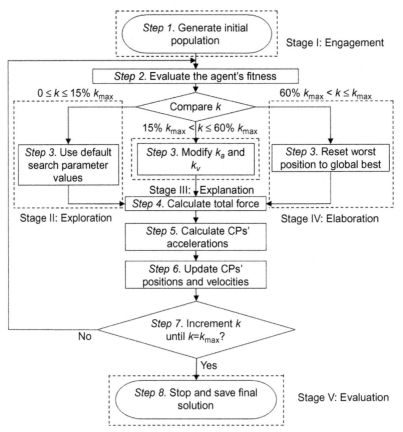

Fig. 3.4 Flowchart of adaptive CSS algorithm. *(From Precup, R.-E., David, R.-C., Petriu, E.M., Preitl, S., Radac, M.-B., 2014. Novel adaptive charged system search algorithm for optimal tuning of fuzzy controllers. Expert Syst. Appl. 41(4, part 1), 1168–1175.)*

Some aspects concerning the stages of this algorithm are presented as follows. The search process in CSS algorithms depends generally on the number of agents N, on the maximum number of iterations k_{max}, and on the parameters $k_a(k)$ and $k_v(k)$. Stage I concerns the generation of the initial population of CPs, that is, the initialization of the q-dimensional search space, of N, the initialization of the iteration index k to $k=0$ and will be incremented at the end of the iteration according to step 7 in Fig. 3.4, the random initialization of CPs' position vector $\mathbf{X}_i$, and the initialization of k_{max}. Since N and k_{max} are constant, the adaptive CSS algorithm presented in this section carries out the adaptation of $k_a(k)$ and $k_v(k)$ to k.

Stage II allows the algorithm to discover the extent of the search space. This stage is characterized by conserving the initial parameters values during the first 20% of k_{max} runs of the search process.

This adaptive CSS algorithm restricts CPs' movements in stage III by the introduction of the modification laws for $k_a(k)$ and $k_v(k)$ in terms of Eq. (3.13) in order to reduce inter-CP distances.

Stage IV is characterized by setting the general position for the optimal value of fitness function and leaving the remaining time to refine the obtained results. The values of all search parameters are frozen, and only the worst CPs' positions are reset to the best values obtained so far after each run.

Stage V focuses on the evaluation of real-world optimization problem's performance for the location of the best position vector obtained during the search process. The obtained solution is mapped onto the real-world optimization problem and tested at this stage.

This adaptive CSS algorithm was employed in step 4 of the design approach dedicated to the simple Takagi-Sugeno PI-FCs presented in Section 1.4. The adaptive CSS algorithm-based solving of the optimization problems required the a priori setting of the algorithm's parameters as follows. The number of CPs was set to $N = 20$. The maximum number of iterations of the search process was set to $k_{max} = 100$. As in the case of the standard version of CSS, each CP has a uniform volume charge density and is considered as a charged sphere with radius $a = 1$. In order to avoid a possible division by zero, the ε parameter in Eq. (3.9) was set to $\varepsilon = 10^{-4}$.

A set of results pertaining to running the adaptive CSS algorithm to minimize the objective functions $J_1, J_2, J_3,$ and J_4 is presented in Table 3.2. The dynamic regimes for the evaluation of the objective functions were specified in Section 1.4. The results given in Table 3.2 were obtained after several restarts of the adaptive CSS algorithm for every objective function, which were imposed as a consequence for the degrees of freedom of the random

Table 3.2 Results for the adaptive CSS-based minimization of J_Θ, $\Theta = 1, \ldots, 4$

J_Θ	$B_{\Delta e}^*$	B_e^*	η^*	β^*	k_c^*	T_i^*	$J_{\Theta min}$
J_1	0.1385	40	0.75	3.1435	0.0044	2.892	390,459
J_2	0.0859	40	0.75	5.0849	0.0034	4.678	22,975.7
J_3	0.0850	39.7528	0.75	5.0866	0.0034	4.6797	2,991,580
J_4	0.0845	40	0.75	5.1529	0.0034	4.7407	155,249

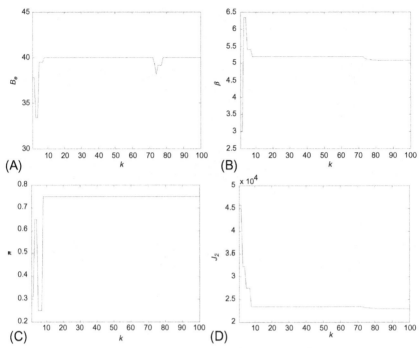

Fig. 3.5 Takagi-Sugeno PI-FC tuning parameters and objective function J_2 evolution vs iteration index in adaptive CSS algorithm running: B_e vs k (A), β vs k (B), η vs k (C), and J_2 vs k (D).

parameters contained by the search process. An in–depth analysis of this aspect is presented in Chapter 5 focusing on the average values of the objective functions along with three performance indices.

In Fig. 3.5, the progress of the search process is tracked for the best CP results. Each graph corresponds to the parameters given by the algorithm together with the values corresponding to the objective function J_2.

Fig. 3.6 displays the movement of all CPs employed by the adaptive CSS algorithm at the end of the first four stages of the algorithm. These positions are expressed as vector solutions ρ to the optimization problem (1.80), and they belong to the search domain D_ρ.

3.3 Fuzzy logic-based adaptive gravitational search algorithms

As shown by Melin et al. (2013), although many nature–inspired optimization algorithms have been developed recently, they are not always able to solve some problems in the best way as there are not yet approaches available

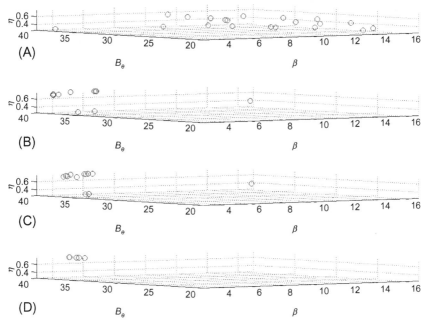

Fig. 3.6 Evolution of vector solution ρ to the optimization problem (1.80) solved by adaptive CSS algorithm in the search domain D_ρ for four values of iteration index k: $k=1$ (A), $k=15$ (B), $k=60$ (C), and $k=100$ (D).

to get the best parameters of the algorithms that can be set at the beginning when using the algorithms. A successful attempt to improve the nature-inspired optimization algorithms is represented by the use of fuzzy logic to adapt parameters and achieve better results than with the initial algorithms, and a general overview is given by Valdez et al. (2014).

Representative results on the fuzzy logic-based adaptation of the parameters of nature-inspired optimization algorithms are briefly discussed as follows. Three measures are used as inputs of a fuzzy system designed by Melin et al. (2013) to adapt two parameters of particle swarm optimization algorithms. The iterations and the diversity of the agents in specific moments of algorithm's execution are used by Olivas et al. (2017) as inputs of the fuzzy system that adapts some parameters of GSA. Type-2 fuzzy logic systems are proposed by Castillo et al. (2018) to adapt some parameters of ant colony optimization (ACO) and GSA, where the abilities to perform the global or local search are controlled by metrics that include the percentage of iterations elapsed and the diversity of the population; the comparison of ACO and GSA is next conducted by Olivas et al. (2018). The parameter ζ specific to the exponential decrease law of the gravitational constant (3.7) in GSA is adapted by Pelusi et al. (2017) using an evolutionary fuzzy system. The parameter ζ and a parameter in the formula that gives the best agents in GSA are next adapted by Pelusi et al. (2018) using fuzzy systems. Mutation is introduced by Kherabadi et al. (2017), and since it should not be applied to all obtained solutions, a fuzzy system has been used to obtain some percentage of masses specific to GSA. A fuzzy logic block is proposed by Precup et al. (2013b) to adapt the gravitational constant $g(k)$ and the parameter ε in Eq. (3.4). Popov's hyperstability analysis is applied as follows to set the parameters of the fuzzy block.

The application of the hyperstability results to GSA's equations given in Eq. (3.3) requires the modification of the state-space equations (3.3) to be expressed as a discrete-time dynamical feedback system in the iteration domain. Introducing the additional variables $b_i^d(k)$:

$$b_i^d(k) = a_i^d(k), \quad d = 1, \ldots, q, \quad i = 1, \ldots, N, \tag{3.14}$$

and replacing the expression of $v_i^d(k+1)$ from the first to the second equation in Eq. (3.3), the modified expression of the state-space equations (3.3) is

$$
\begin{aligned}
v_i^d(k+1) &= \rho_i v_i^d(k) + a_i^d(k), \\
x_i^d(k+1) &= \rho_i v_i^d(k) + x_i^d(k) + b_i^d(k).
\end{aligned}
\tag{3.15}
$$

A feedforward discrete-time linear time-invariant (LTI) system is inserted. This LTI system is characterized by the multi-input multi-output (MIMO) state-space model

$$
\begin{aligned}
\mathbf{x}_i^{\mathrm{LTI}}(k+1) &= \mathbf{A}_i^{\mathrm{LTI}} \mathbf{x}_i^{\mathrm{LTI}}(k) + \mathbf{B}_i^{\mathrm{LTI}} \mathbf{u}_i^{\mathrm{LTI}}(k), \\
\mathbf{v}_i^{\mathrm{LTI}}(k) &= \mathbf{C}_i^{\mathrm{LTI}} \mathbf{x}_i^{\mathrm{LTI}}(k) + \mathbf{J}_i^{\mathrm{LTI}} \mathbf{u}_i^{\mathrm{LTI}}(k),
\end{aligned}
\tag{3.16}
$$

with the input vector $\mathbf{u}_i^{\mathrm{LTI}}$, state-space vector $\mathbf{x}_i^{\mathrm{LTI}}$, and output vector $\mathbf{v}_i^{\mathrm{LTI}}$

$$
\begin{aligned}
\mathbf{u}_i^{\mathrm{LTI}} &= \begin{bmatrix} a_i^1 & a_i^2 & \cdots & a_i^q & a_i^1 & a_i^2 & \cdots & a_i^q \end{bmatrix}^T, \\
\mathbf{x}_i^{\mathrm{LTI}} = \mathbf{v}_i^{\mathrm{LTI}} &= \begin{bmatrix} v_i^1 & v_i^2 & \cdots & v_i^q & x_i^1 & x_i^2 & \cdots & x_i^q \end{bmatrix}^T,
\end{aligned}
\tag{3.17}
$$

and the matrices

$$
\mathbf{A}_i^{\mathrm{LTI}} = \begin{bmatrix} \rho_i \operatorname{diag}(1, 1, \ldots, 1) & \mathbf{0}_{q,q} \\ \rho_i \operatorname{diag}(1, 1, \ldots, 1) & \operatorname{diag}(1, 1, \ldots, 1) \end{bmatrix}, \quad \operatorname{diag}(1, 1, \ldots, 1) \in \mathbf{R}^{q \times q},
$$

$$
\mathbf{B}_i^{\mathrm{LTI}} = \mathbf{C}_i^{\mathrm{LTI}} = \operatorname{diag}(1, 1, \ldots, 1) \in \mathbf{R}^{2q \times 2q}, \quad \mathbf{J}_i^{\mathrm{LTI}} = \mathbf{0}_{2q, 2q} \in \mathbf{R}^{2q \times 2q}.
\tag{3.18}
$$

As shown in Eq. (3.18), the LTI system has an equal number of $2q$ inputs and outputs, required by Popov's hyperstability analysis results in the MIMO case considered here. Therefore, the introduction of the variables $b_i^d(k)$ in Eq. (3.14) is justified.

Eqs. (3.3)–(3.6) are organized in terms of the nonlinear (NL) feedback block with the output vector $\mathbf{w}_i^{\mathrm{LTI}}(k)$

$$\mathbf{w}_i^{\mathrm{LTI}}(k) = -\left[a_i^1(k)\ a_i^2(k)\ \cdots\ a_i^q(k)\ a_i^1(k)\ a_i^2(k)\ \cdots\ a_i^q(k)\right]^T, \quad (3.19)$$

which depends on $\mathbf{v}_i^{\mathrm{LTI}}(k)$.

The discrete transfer function matrix $\mathbf{H}(z)$ of LTI block is

$$\mathbf{H}(z) = \mathbf{J}_i^{\mathrm{LTI}} + \mathbf{C}_i^{\mathrm{LTI}}\left(z\mathbf{I}_{2q,2q} - \mathbf{A}_i^{\mathrm{LTI}}\right)^{-1}\mathbf{B}_i^{\mathrm{LTI}}, \quad \mathbf{I}_{2q,2q} = \mathrm{diag}(1, 1, ..., 1)$$
$$\in \mathbf{R}^{2q \times 2q}, \qquad\qquad\qquad (3.20)$$

where the expression of the matrix whose inverse is needed is

$$z\mathbf{I}_{2q,2q} - \mathbf{A}_i^{\mathrm{LTI}} = \begin{bmatrix} (z - \rho_i)\,\mathrm{diag}(1, 1, ..., 1) & \mathbf{0}_{q,q} \\ -\rho_i\,\mathrm{diag}(1, 1, ..., 1) & (z - 1)\,\mathrm{diag}(1, 1, ..., 1) \end{bmatrix} = \begin{bmatrix} \mathbf{A}(z) & \mathbf{B} \\ \mathbf{C} & \mathbf{D}(z) \end{bmatrix},$$

$$\mathbf{A}(z) = (z - \rho_i)\,\mathrm{diag}(1, 1, ..., 1) \in R^{q \times q}, \quad \mathbf{B} = \mathbf{0}_{q,q} \in \mathbf{R}^{q \times q}, \quad \mathbf{C} = -\rho_i\,\mathrm{diag}(1, 1, ..., 1) \in \mathbf{R}^{q \times q},$$

$$\mathbf{D}(z) = (z - 1)\,\mathrm{diag}(1, 1, ..., 1) \in \mathbf{R}^{q \times q}. \qquad\qquad (3.21)$$

The block-wise matrix inversion formula according to (Bernstein, 2005)

$$\begin{bmatrix} \mathbf{A}(z) & \mathbf{B} \\ \mathbf{C} & \mathbf{D}(z) \end{bmatrix}^{-1}$$

$$= \begin{bmatrix} \mathbf{A}^{-1}(z) + \mathbf{A}^{-1}(z)\mathbf{B}\left[\mathbf{D}(z) - \mathbf{C}\mathbf{A}^{-1}(z)\mathbf{B}\right]^{-1}\mathbf{C}\mathbf{A}^{-1}(z) & -\mathbf{A}^{-1}\mathbf{B}\left[\mathbf{D}(z) - \mathbf{C}\mathbf{A}^{-1}(z)\mathbf{B}\right]^{-1} \\ -\left[\mathbf{D}(z) - \mathbf{C}\mathbf{A}^{-1}(z)\mathbf{B}\right]^{-1}\mathbf{C}\mathbf{A}^{-1}(z) & \left[\mathbf{D}(z) - \mathbf{C}\mathbf{A}^{-1}(z)\mathbf{B}\right]^{-1} \end{bmatrix} \qquad (3.22)$$

applied for $\mathbf{B}$ in Eq. (3.21) leads to

$$\left(z\mathbf{I}_{2q,2q} - \mathbf{A}_i^{\mathrm{LTI}}\right)^{-1} = \begin{bmatrix} \mathbf{A}(z) & \mathbf{B} \\ \mathbf{C} & \mathbf{D}(z) \end{bmatrix}^{-1} = \begin{bmatrix} \mathbf{A}^{-1}(z) & \mathbf{0}_{q,q} \\ -\mathbf{D}^{-1}(z)\mathbf{C}\mathbf{A}^{-1}(z) & \mathbf{D}^{-1}(z) \end{bmatrix}$$

$$= \begin{bmatrix} \dfrac{1}{z - \rho_i}\,\mathrm{diag}(1, 1, ..., 1) & \mathbf{0}_{q,q} \\ \dfrac{1}{(z - \rho_i)(z - 1)}\,\mathrm{diag}(1, 1, ..., 1) & \dfrac{1}{z - 1}\,\mathrm{diag}(1, 1, ..., 1) \end{bmatrix}. \qquad (3.23)$$

Replacing in Eq. (3.20) the expression of $(z\,\mathbf{I}_{2q,2q} - \mathbf{A}_i^{\mathrm{LTI}})^{-1}$ taken from Eq. (3.23), the expression of the matrix $\mathbf{H}(z)$ is

$$\mathbf{H}(z) = \begin{bmatrix} \dfrac{1}{z - \rho_i}\,\mathrm{diag}(1, 1, ..., 1) & \mathbf{0}_{q,q} \\ \dfrac{1}{(z - \rho_i)(z - 1)}\,\mathrm{diag}(1, 1, ..., 1) & \dfrac{1}{z - 1}\,\mathrm{diag}(1, 1, ..., 1) \end{bmatrix}. \qquad (3.24)$$

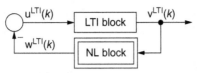

Fig. 3.7 Equivalent feedback structure of GSA's operating equations in the iteration domain.

Using the common vector variables, the LTI and NL blocks are connected in terms of the feedback structure presented in Fig. 3.7 that is used in GSA's convergence analysis. In this regard GSA's convergence is equivalent to the stability of the NL MIMO feedback system characterized by Eqs. (3.16), (3.19), (3.24). The system illustrated in Fig. 3.7 is also an autonomous system in the discrete-time iteration domain.

A convergence analysis theorem is given by Precup et al. (2013b) and proved using Popov's hyperstability results for NL discrete-time MIMO systems according to the formulation presented by Landau (1979). This theorem states that the sufficient convergence condition for GSA is

$$\sum_{k=k_0}^{k_1} \left(\mathbf{w}_i^{LTI}(k) \right)^T \mathbf{v}_i^{LTI}(k) \geq -\mu_0^2, \quad \forall k_1 \geq k_0 \geq 0, \quad \mu_0 = \text{const}, \quad \mu_0 \neq 0,$$

(3.25)

which points out the Popov sum.

The search process in GSA depends on the parameters of GSA, namely the number of agents N, the maximum number of iterations k_{max}, the gravitational constant $g(k)$, and the parameter ε which have a positive impact on algorithm's results. Since the number of agents and iterations are constant, the adaptive GSA presented in this section employs the fuzzy logic-based adaptation of $g(k)$ and ε to k (using $\varepsilon(k)$ instead of ε in Eq. (3.4)) on the basis of a single input–two output (SITO) fuzzy block. This approach is different than that proposed by Askari and Zahiri (2012), where a fuzzy logic block adapts the number of effective agents and $g(k)$.

The SITO fuzzy block ensures the improvement in the search accuracy by the linguistic description of the effects of GSA parameters along its search process. The SITO fuzzy block is a Takagi–Sugeno fuzzy system without dynamics, characterized by the structure and input membership functions presented in Fig. 3.8.

The rule base of the SITO fuzzy block consists of three rules, R^1, R^2, and R^3

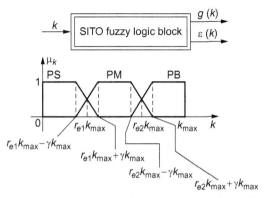

Fig. 3.8 Structure and input membership functions of SITO fuzzy block.

R^1: IF k IS PS THEN $g(k) = g_0(1 - \psi k/k_{max})$, $\varepsilon(k) = \varepsilon_0$,

R^2: IF k IS PM THEN $g(k) = g_0 \exp(-\zeta k/k_{max})$, $\varepsilon(k) = \varepsilon_0 - \dfrac{\varepsilon_0(k - r_{e1}k_{max})}{(1 - r_{e1})k_{max}}$,

R^3: IF k IS PB THEN $g(k) = g(k-1)$, $\varepsilon(k) = \varepsilon_0 - \dfrac{\varepsilon_0(k - r_{e1}k_{max})}{(1 - r_{e1})k_{max}}$.

$$(3.26)$$

Rule R^1 allows the algorithm to discover the extent of the search space, that is, R^1 contributes to carrying out the exploration. This rule is characterized by a linear decrease of $g(k)$ according to Eq. (3.2) during the first $r_{e1}k_{max}$ runs of the search process, where the parameter r_{e1} indicates the ratio of exploration runs, $0 < r_{e1} < 1$.

Rule R^2 restricts agents' movements by the introduction of a more aggressive depreciation schedule of $g(k)$ in terms of Eq. (3.7) and by a linear depreciation of ε in order to reduce the inter-agent distances, that is, R^2 contributes to carrying out the explanation. This rule ensures the exponential depreciation of $g(k)$ according to Eq. (3.7) and the reduction of the parameter ε starting with the preset value ε_0 during the next $r_{e2}k_{max}$ runs in the search process of the adaptive GSA, where the parameter r_{e2} indicates the ratio of explanation runs, $0 < r_{e2} < 1$.

Rule R^3 sets the general position for the optimal value of the objective function and leaves the remaining time to refine the obtained results. The value of $g(k)$ stops decaying and only ε continues the depreciation process. Worst agents' positions are reset to the best values obtained so far after each run. This rule corresponds to the last $k_{max}(1 - r_{e1} - r_{e2})$ runs in the search process of the adaptive GSA.

The SUM and PROD operators are used in the inference engine of the SITO fuzzy block. The defuzzification is done by the weighted sum method. Therefore, this SITO fuzzy block behaves like a relatively simple bumpless interpolator between two depreciation laws of the gravitational constant.

The fuzzy logic-based adaptive GSA consists of the steps A, B, and C illustrated in Fig. 3.9. An evaluation of the real-world optimization problem's performance is conducted in step C for the location of the best position vector obtained during steps A and B, and the obtained solution is next validated and mapped onto the real-world optimization problem. The parameters of the input membership functions of the TISO fuzzy block are set as follows in order to ensure a trade-off to convergence speed and number of evaluations of the objective function (Precup et al., 2014b):

$$r_{e1} = 0.15, \quad r_{e2} = 0.6, \quad \gamma = 0.02. \tag{3.27}$$

Table 3.3 illustrates the values of the optimal controller tuning parameters and the minimum values of the objective functions J_1, J_2, J_3, and J_4 related to their minimization by the fuzzy logic-based adaptive GSA described in this section. The dynamic regimes for the evaluation of the objective functions were specified in Section 1.4. The data presented in this

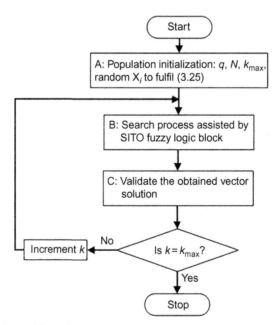

Fig. 3.9 Flowchart of fuzzy logic-based adaptive GSA.

Table 3.3 Results for the fuzzy logic-based adaptive GSA-based minimization of J_Θ, $\Theta = 1, \ldots, 4$

J_Θ	$B_{\Delta e}^*$	B_e^*	η^*	β^*	k_c^*	T_i^*	$J_{\Theta min}$
J_1	0.1386	40	0.743	3.143	0.004379	2.8917	390,417
J_2	0.0843	39.4214	0.75	5.0872	0.0034	4.6802	23,042.5
J_3	0.0851	40	0.745	5.1118	0.0034	4.70288	3,025,701
J_4	0.0856	39.9958	0.75	5.0849	0.0034	4.678	2,984,880

table was obtained after several reruns of the algorithm that were required in order to deal with the arbitrary characteristic of the fuzzy logic–based adaptive GSA. The parameters of the fuzzy logic–based adaptive GSA were set to the same values as those of the adaptive GSA given in Section 3.1: the number of used agents $N = 20$ was set, the maximum number of iterations was set to $k_{max} = 100$. The ψ, ζ, and g_0 parameters in Eqs. (3.2) and (3.7) were set as $\psi = 0.5$, $\zeta = 8.5$, and $g_0 = 100$, and the ε_0 parameter in Eq. (3.26) was set to $\varepsilon_0 = 10^{-4}$.

The results presented in Table 3.3 indicate a small performance improvement for two of the objective functions compared to the results specific to the adaptive GSA presented in Table 3.1. The improvement is reflected in the reduction of the values of J_1 and J_3 after running the fuzzy logic–based GSA because smaller values of η^* were obtained compared to those obtained after running the adaptive GSA. Therefore, the evolutions of the parameters and all agents' positions are similar to those illustrated in Figs. 3.2 and 3.3, respectively.

References

Askari, H., Zahiri, S.-H., 2012. Decision function estimation using intelligent gravitational search algorithm. Int. J. Mach. Learn. Cybern. 3 (2), 163–172.

Balci, S., Cakiroglus, J., Tekkayas, C., 2006. Engagement, exploration, explanation, extension, and evaluation (5E) learning cycle and conceptual change text as learning tools. Biochem. Mol. Biol. Educ. 34 (3), 199–203.

Bernstein, D.S., 2005. Matrix Mathematics. Princeton University Press, Princeton, NJ, USA.

Bybee, R.W. (Ed.), 2002. Learning Science and the Science of Learning: Science Educators' Essay Collection. NSTA Press, Arlington, VA.

Castillo, O., Olivas, F., Valdez, F., 2018. Dynamic parameter adaptation using interval type-2 fuzzy logic in bio-inspired optimization methods. In: Abraham, A., Haqiq, A., Muda, A.K., Gandhi, N. (Eds.), Innovations in Bio-Inspired Computing and Applications. IBICA 2017. In: Advances in Intelligent Systems and Computing, vol. 375. Springer, Cham, pp. 1–12.

David, R.-C., Precup, R.-E., Petriu, E.M., Radac, M.-B., Preitl, S., 2013. Gravitational search algorithm-based design of fuzzy control systems with a reduced parametric sensitivity. Inform. Sci. 247, 154–173.

Kaveh, A., Talatahari, S., 2010a. A novel heuristic optimization method: charged system search. Acta Mech. 213, 267–289.

Kaveh, A., Talatahari, S., 2010b. Optimal design of truss structures via the charged system search algorithm. Struct. Multidiscip. Optim. 37 (6), 893–911.

Kaveh, A., Talatahari, S., 2010c. A charged system search with a fly to boundary method for discrete optimum design of truss structures. Asian J. Civ. Eng. (Build. Hous.) 11 (3), 277–293.

Kherabadi, H.A., Mood, S.E., Javidi, M.M., 2017. Mutation: a new operator in gravitational search algorithm using fuzzy controller. Cybern. Inf. Technol. 17 (1), 72–86.

Landau, Y.-D., 1979. Adaptive Control: The Model Reference Approach. Marcel Dekker, New York, USA.

Liu, K., Tan, Y., He, X., 2010. An adaptive staged PSO based on particles' search capabilities. In: Tan, Y., Shi, Y., Tan, K.C. (Eds.), Advances in Swarm Intelligence. In: Lecture Notes in Computer Science, vol. 6145. Springer-Verlag, Berlin, Heidelberg, pp. 52–59.

Mahmoodabadi, M.J., Danesh, N., 2018. Gravitational search algorithm-based fuzzy control for a nonlinear ball and beam system. J. Control Decis. 5 (3), 229–240.

Melin, P., Olivas, F., Castillo, O., Valdez, F., Soria, J., Valdez, M., 2013. Optimal design of fuzzy classification systems using PSO with dynamic parameter adaptation through fuzzy logic. Expert Syst. Appl. 40 (8), 3196–3206.

Olivas, F., Valdez, F., Castillo, O., 2017. Gravitational search algorithm with parameter adaptation through a fuzzy logic system. In: Melin, P., Castillo, O., Kacprzyk, J. (Eds.), Nature-Inspired Design of Hybrid Intelligent Systems. In: Studies in Computational Intelligence, vol. 667. Springer, Cham, pp. 391–405.

Olivas, F., Valdez, F., Castillo, O., 2018. Comparison of bio-inspired methods with parameter adaptation through interval type-2 fuzzy logic. In: Castillo, O., Melin, P., Kacprzyk, J. (Eds.), Fuzzy Logic Augmentation of Neural and Optimization Algorithms: Theoretical Aspects and Real Applications. In: Studies in Computational Intelligence, vol. 749. Springer, Cham, pp. 39–53.

Pelusi, D., Mascella, R., Tallini, L., 2017. Revised gravitational search algorithms based on evolutionary-fuzzy systems. Algorithms 11 (736), 1–18. https://doi.org/10.3390/en11040736.

Pelusi, D., Mascella, R., Tallini, L., 2018. A fuzzy gravitational search algorithm to design optimal IIR filters. Algorithms 10 (44), 1–19. https://doi.org/10.3390/a10020044.

Precup, R.-E., David, R.-C., Petriu, E.M., Preitl, S., Radac, M.-B., 2012a. Novel adaptive gravitational search algorithm for fuzzy controlled servo systems. IEEE Trans. Ind. Inf. 8 (4), 791–800.

Precup, R.-E., David, R.-C., Petriu, E.M., Preitl, S., Radac, M.-B., 2012b. Charged system search algorithms for optimal tuning of PI controllers. IFAC Proc. Vol. 45 (4), 115–120.

Precup, R.-E., David, R.-C., Petriu, E.M., Preitl, S., Radac, M.-B., 2013a. Experiments in fuzzy controller tuning based on an adaptive gravitational search algorithm. Proc. Romanian Acad., Series A: Math. Phys. Tech. Sci. Inform. Sci. 14 (4), 360–367.

Precup, R.-E., David, R.-C., Petriu, E.M., Preitl, S., Radac, M.-B., 2013b. Fuzzy logic-based adaptive gravitational search algorithm for optimal tuning of fuzzy controlled servo systems. IET Control Theory Appl. 7 (1), 99–107.

Precup, R.-E., David, R.-C., Petriu, E.M., Preitl, S., Radac, M.-B., 2014a. Novel adaptive charged system search algorithm for optimal tuning of fuzzy controllers. Expert Syst. Appl. 41 (4, part 1), 1168–1175.

Precup, R.-E., David, R.-C., Petriu, E.M., Radac, M.-B., Preitl, S., 2014b. Adaptive GSA-based optimal tuning of PI controlled servo systems with reduced process parametric sensitivity, robust stability and controller robustness. IEEE Trans. Cybern. 44 (11), 1997–2009.

Precup, R.-E., David, R.-C., Petriu, E.M., Radac, M.-B., Preitl, S., Fodor, J., 2013c. Evolutionary optimization-based tuning of low-cost fuzzy controllers for servo systems. Knowl.-Based Syst. 38, 74–84.

Precup, R.-E., Sabau, M.-C., Petriu, E.M., 2015. Nature-inspired optimal tuning of input membership functions of Takagi-Sugeno-Kang fuzzy models for anti-lock braking systems. Appl. Soft Comput. 27, 575–589.

Rashedi, E., Nezamabadi-pour, H., Saryazdi, S., 2009. GSA: a gravitational search algorithm. Inform. Sci. 179 (13), 2232–2248.

Rashedi, E., Nezamabadi-pour, H., Saryazdi, S., 2010. BGSA: binary gravitational search algorithm. Nat. Comput. 9 (3), 727–745.

Valdez, F., Melin, P., Castillo, O., 2014. A survey on nature-inspired optimization algorithms with fuzzy logic for dynamic parameter adaptation. Expert Syst. Appl. 41 (14), 6459–6466.

CHAPTER 4

Hybrid nature-inspired algorithms for the optimal tuning of fuzzy controllers

Contents

4.1 Hybrid particle swarm optimization-gravitational search algorithms   103
4.2 Hybrid gray wolf optimization-particle swarm optimization algorithms   108
References   113

Abstract

This chapter first describes the hybridization of particle swarm optimization (PSO) and gravitational search algorithms (GSAs) and later introduces a gray wolf optimization (GWO) using PSO's search mechanism. The operating algorithm of a hybrid PSOGSA and GWOPSO are presented. Combining two nature-inspired algorithms is necessary in order to reduce one's search process drawbacks by using the other's fortes. In the case of the PSOGSA, the PSO's exploitation capabilities and GSA's exploration abilities are combined to avoid getting trapped in local minima situations, while in GWOPSO's case, the exploitation advantages of PSO are employed in order to speed the GWO's convergence. The hybrid algorithms are inserted in the design approach dedicated to the optimal tuning of simple Takagi-Sugeno proportional-integral fuzzy controllers for the position control of servo systems. Four optimization problems are solved and some results concerning the hybrid PSOGSA's behavior are presented.

Keywords: Exploration, Exploitation, Hybrid particle swarm optimization-gravitational search algorithms, Hybrid gray wolf optimization-particle swarm optimization, Takagi-Sugeno proportional-integral fuzzy controllers

4.1 Hybrid particle swarm optimization-gravitational search algorithms

The hybridization of nature–inspired algorithms evolved as a solution necessary in overcoming certain shortcomings observed during the use of classical algorithms. A hybridization of particle swarm optimization (PSO) and gravitational search algorithm (GSA) algorithms has been proposed by Mirjalili and Hashim (2010) with the objective of obtaining an improved search technique, which aims to incorporate the advances of both algorithms. In order to achieve this goal, the ability of social thinking in PSO is interrelated with the local search capability of GSA. This hybrid algorithm is applied in several versions in various domains: tuning of fuzzy controllers to ensure a reduced process parametric sensitivity (David et al., 2012), flow-based anomaly detection (Jadidi et al., 2013), emission load dispatch (Jiang et al., 2014), economic and emission dispatch (Dubey et al., 2013), landslide displacement (Lian et al., 2013), supply chain (Pei et al., 2014), robotic path planning (Purcaru et al., 2013), speech enhancement (Prajna et al., 2015), power system control (Khadanga and Satapathy, 2015), data mining (Singh et al., 2017), and reinforced concrete structural frame (Chutani and Singh, 2018).

The operating mechanism of the PSO algorithm, based on the use of swarm particles, also called agents and presented in Chapter 2, is employed in the framework of the hybrid PSOGSA. The agents continue to be characterized by vectors $\mathbf{X}_i$ (the particle position vector) and $\mathbf{V}_i$ (the particle velocity vector) (Kennedy and Eberhart, 1995a, b).

$$
\begin{aligned}
\mathbf{X}_i &= \left[x_i^1 \ \cdots \ x_i^d \ \cdots \ x_i^q \right]^T, \\
\mathbf{V}_i &= \left[v_i^1 \ \cdots \ v_i^d \ \cdots \ v_i^q \right]^T,
\end{aligned}
\tag{4.1}
$$

where i, $i=1,\ldots,N$, is the index of current agent in the swarm, N represents the size of the swarm, and q represents the dimension of the search space. The particle position vector will have a $q=3$ dimension in order to suit Eq. (1.89) for optimization problems (1.79)–(1.82). Let $\mathbf{P}_{g,\,Best}$ be the best swarm position vector

$$
\mathbf{P}_{g,Best} = \left[p_g^1 \ \cdots \ p_g^d \ \cdots \ p_g^q \right]^T.
\tag{4.2}
$$

The vector $\mathbf{P}_{g,\,Best}$ is used as in the case of PSO, and is updated according to Eq. (2.6). The computation of the initial values of $\mathbf{P}_{g,\,Best}$ will be presented and carried out in the first step of the hybrid PSOGSA.

The integration of PSO's exploitation capabilities and GSA's exploration abilities are highlighted during the agent's velocities and positions update according to

$$
v_i^d(k+1) = \begin{cases} w(k)\, v_i^d(k) + c_1 r_1 \left[p_g^d(k) - x_i^d(k) \right] + c_2\, r_2 a_i^d(k) & \text{if } m_i(k) > 0, \\ w(k)\, v_i^d(k) + c_1 r_1 \left[p_g^d(k) - x_i^d(k) \right] & \text{otherwise,} \end{cases}
$$
$$
x_i^d(k+1) = x_i^d(k) + v_i^d(k+1), \quad d=1,\ldots,q, \quad i=1,\ldots,N,
\tag{4.3}
$$

where r_1 and r_2 are uniformly distributed random variables, $0 \leq r_1 \leq 1$, $0 \leq r_2 \leq 1$, c_1 and c_2, $c_1 \geq 0$, $c_2 \geq 0$, represent weighting factors; the parameter $w(k)$ stands for the inertia weight and $a_i^d(k)$ is the acceleration expressed in Eq. (2.15).

The hybrid PSOGSA algorithm consists of the following steps, presented also by David et al. (2012) and pointed out in Fig. 4.1:

Step 1. Generate the initial population of agents, that is, initialize the q-dimensional search space, the number of agents N, set the iteration index $k = 0$, set the search process iteration limit $k_{\max}$, the weighting factors c_1, c_2, the inertia weight parameter $w(k)$ according to Eq. (2.4), and initialize randomly the agents' position vector $\mathbf{X}_i(0)$. Define the gravitational constant decrease law.

Step 2. Evaluate the agents' fitness using the objective functions defined in Eqs. (1.79)–(1.82).

Step 3. Compare the performance of each particle to the best global performance, and eventually update the best swarm position vector $\mathbf{P}_{g, \, Best}$ according to Eq. (2.6).

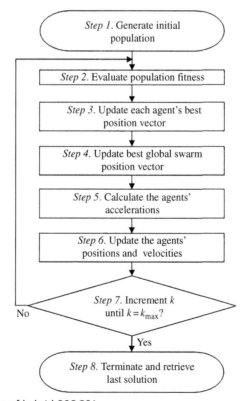

Fig. 4.1 Flowchart of hybrid PSOGSA.

Step 4. Calculate the agents' accelerations $a_i^d(k)$ according to Eq. (2.15).
Step 5. Update the agents' velocities $v_i^d(k+1)$ and positions $x_i^d(k+1)$ using Eq. (4.3) for all agents, that is, for $i = 1, \dots, N$.
Step 6. Continue incrementing k and go to step 2 until the maximum number of iterations is reached, that is, until $k = k_{max}$.
Step 7. Terminate and retrieve the final solution in the vector $\mathbf{X}_i$ obtained so far.

This hybrid PSOGSA was involved in step 4 of the design approach dedicated to the simple Takagi-Sugeno proportional-integral fuzzy controllers (PI-FCs) presented in Section 1.4. Employing the hybrid PSOGSA requires the definition of all parameters of algorithms specified in step 1, and these values are presented as follows. The number of agents was set to $N = 20$. The maximum number of iterations of the search process was set to $k_{max} = 100$. As in the case of GSA, the decrease law (2.9) of the gravitational constant was used, with the initial value g_0 set to $g_0 = 100$ and $\zeta = 8.5$. The ε value in Eq. (2.12), introduced to avoid possible divisions by zero, was set to $\varepsilon = 10^{-4}$. In the PSO part, the weighting parameters were set to $c_1 = c_2 = 0.3$ in order to ensure a good balance between exploration and exploitation characteristics. The inertia weight parameters outlined in Eq. (2.4) were set to $w_{max} = 0.9$ and $w_{min} = 0.5$.

Table 4.1 illustrates the values of the optimal controller tuning parameters and the minimum values of the objective functions J_1, J_2, J_3, and J_4 related to their hybrid PSOGSA-based minimization. The dynamic regimes for the evaluation of the objective functions were specified in Section 1.4. The data presented in this table was obtained after several reruns of the algorithm that were required in order to deal with the arbitrary characteristic of the hybrid PSOGSA. A more detailed analysis based on the average values of the objective functions, together with other algorithm performance indices is presented in Chapter 5.

In order to have a better representation of the search process, Fig. 4.2 illustrates the evolution of parameters defined by the search domain in the case of objective function J_2.

Table 4.1 Results for the hybrid PSOGSA-based minimization of J_Θ, $\Theta = 1, \dots, 4$

J_Θ	$B_{\Delta e}^*$	B_e^*	η^*	β^*	k_c^*	T_i^*	$J_{\Theta min}$
J_1	0.1362	40	0.75	3.19747	0.004342	2.94168	390,671
J_2	0.0809	37.8291	0.75	5.0852	0.003443	4.67842	23,119.3
J_3	0.0756	35.561	0.8835	5.1202	0.0034	4.71054	2,903,740
J_4	0.0856	40	0.75	5.08538	0.003443	4.67855	153,530

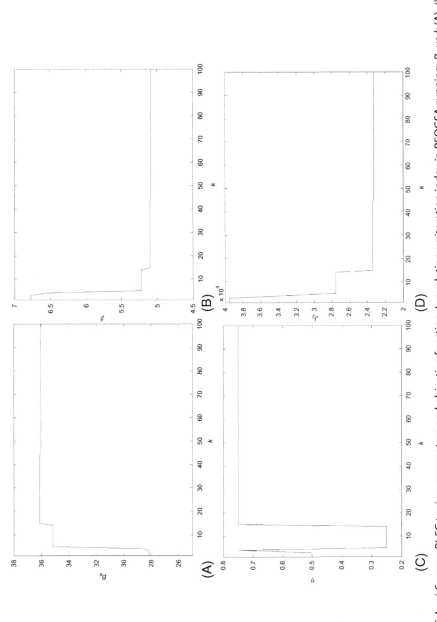

Fig. 4.2 Takagi-Sugeno PI-FC tuning parameters and objective function J_2 evolution vs iteration index in PSOGSA running: B_e vs k (A), β vs k (B), η vs k (C), and J_2 vs k (D).

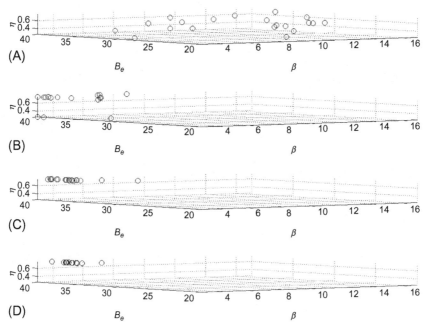

Fig. 4.3 Evolution of vector solution ρ to the optimization problem (1.80) solved by PWOGSA in the search domain D_ρ for four values of iteration index k: $k=1$ (A), $k=15$ (B), $k=60$ (C), and $k=100$ (D).

Fig. 4.3 illustrates the evolution of all agents' positions for the hybrid PSOGSA after the first four stages of the search process. These positions are expressed as vector solutions ρ to the optimization problem (1.80), and they belong to the search domain D_ρ.

4.2 Hybrid gray wolf optimization-particle swarm optimization algorithms

The gray wolf optimization (GWO) algorithm, discussed in Section 2.4, is a fairly recent addition to the family of metaheuristics. Its inspiration for a search process based on the behavior and social hierarchy of wolf packs leaves it dependent on the interactions between the top three agents, called alpha, beta, and omega, while the remainder of the pack, called gammas, has a diminished role in the social interactions of the pack. The search stages described by searching and attacking the prey correspond to the exploration and exploitation abilities of the search process.

By abstracting this information to the described GWO search process, it can be observed that the exploitation abilities of GWO are limited to the knowledge of the top three performers. Although this is an advantage in the exploration phase, as the effort of the search is focused on the direction given by the best agents, for exploitation this solution has its drawbacks as it disregards the effort of the remaining members of the pack.

In recent research there are several solutions proposed to overcome this limitation by integration of differential evolution (Jitkongchuen, 2015), genetic algorithms (Tawhid and Ali, 2017), elite opposition-based learning and simplex method (Zhang et al., 2017), and sine cosine algorithm (Singh and Singh, 2017a). However, PSO is one of the most versatile hybridization variants due to its proven stability and extensive studies that were based on it. The PSO's search process is one of the most documented evolutionary heuristic, given its capability of finding global solutions with a reduced implementation cost and increased convergence speed.

Following the results presented by Singh and Singh (2017b), a hybrid GWOPSO algorithm's search process is described based on the functional scheme of GWO in which the top three agents' position update relationship (2.35) is replaced with.

$$d_l^{if}(k) = |c_l^f(k)x^{lf}(k) - \omega x_i^f(k)|, \quad i = 1, \ldots, N, \quad l \in \{\alpha, \beta, \delta\}, \qquad (4.4)$$

where PSO's inertia weight parameter ω (with the notation $w(k)$ in Section 2.1) is introduced to regulate the agents' exploration and exploitation capabilities.

The GWO's position update law (2.35) is replaced with

$$\mathbf{X}_i(k+1) = \mathbf{X}_i(k) + \omega \{ c_\alpha r_\alpha [\mathbf{X}^\alpha(k+1) - \mathbf{X}_i(k)] + c_\beta r_\beta [\mathbf{X}^\beta(k+1) - \mathbf{X}_i(k)]$$
$$+ c_\delta r_\delta [\mathbf{X}^\delta(k+1) - \mathbf{X}_i(k)] \}, \quad i = 1, \ldots, N, \qquad (4.5)$$

where a new component c_l, $l \in \{\alpha, \beta, \gamma\}$ is introduced for updating agents' positions, along with ω and a random parameter $0 < r_l < 1$, $l \in \{\alpha, \beta, \delta\}$. The position update law (4.5) continues to use the current agent's position and the top three performers in present iteration as inputs, to determine the position for the next iteration. However, compared to the standard form of GWO, in GWOPSO this position update is influenced by a set of parameters that influence the exploration and exploitation capabilities according to the designer's needs.

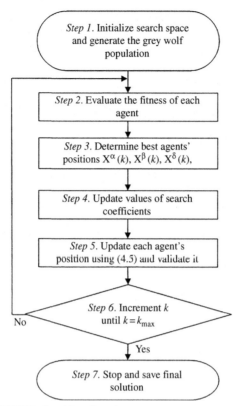

Fig. 4.4 Hybrid GWOPSO algorithm flowchart.

The hybrid GWOPSO algorithm involves the following steps, and shown in Fig. 4.4:

Step 1. Generate the initial random gray wolf population, represented by N agents' positions in the q-dimensional search space, initialize the iteration index to $k=0$, and set the maximum number of iterations to k_{max}.

Step 2. Evaluate the performance of each member of the population of agents by simulations and/or experiments conducted on the fuzzy control system. The evaluation leads to the objective functions (1.79)–(1.82) value by mapping the GWOPSO algorithm onto the optimization problems in terms of Eq. (2.36).

Step 3. Identify the first three best solutions obtained so far, that is, $\mathbf{X}^{\alpha}(k)$, $\mathbf{X}^{\beta}(k)$, and $\mathbf{X}^{\delta}(k)$, using Eq. (2.28).

Step 4. Calculate the search coefficients using Eqs. (2.30) and (2.31).

Step 5. Move the agents to their new positions by computing $\mathbf{X}_i(k+1)$ in terms of Eqs. (4.4) and (4.5) and validate the new positions for condition (1.93).

Step 6. Increment the iteration index k and continue with step 2 until k_{max} is reached.

The GWOPSO algorithm presented here was applied in the optimal tuning of simple Takagi-Sugeno PI-FCs inline with the design method described in Section 1.4. Having the search domain D_ρ in Eq. (1.92) with dimension $q=3$, the chosen parameters for population size $N=20$ and maximum number of iterations $k_{max}=100$, the search parameters were defined as

$$\omega = 0.5(1 + r_\omega), \quad 0 < r_\omega < 1, \quad c_l = 0.5, \quad l \in \{\alpha, \beta, \delta\}, \quad (4.6)$$

where r_ω is a uniformly distributed random number, in order to get the best trade-off to exploration and exploitation capabilities.

The results obtained for the objective functions J_Θ, $\Theta = 1, \dots, 4$, are presented in Table 4.2. As a downside of having heuristic components in the algorithm structure, the search process needs to be rerun before a solution can be presented as final. A more detailed discussion regarding this aspect will be presented in Chapter 5 based on average values of the objective functions and several performance indices.

Fig. 4.5 provides a better understating of GWOPSO algorithm's search process, in which the evolution of the controller parameters and best value of the objective function J_2 are traced throughout the iterations. An evolutionary display throughout the search process for all GWOPSO algorithm's vector solutions ρ to the optimization problem (1.80) belonging to the search domain D_ρ is presented in Fig. 4.6.

Table 4.2 Results for the GWOPSO-based minimization of J_Θ, $\Theta = 1, \dots, 4$

J_Θ	$B^*_{\Delta e}$	B^*_e	η^*	β^*	k^*_c	T^*_i	$J_{\Theta min}$
J_1	0.1368	40	0.75	3.139	0.004382	2.88791	390,460
J_2	0.0855	40	0.75	5.0852	0.003443	4.67841	22,980.2
J_3	0.0855	40	0.75	5.0849	0.003443	4.67811	2,984,850
J_4	0.0844	40	0.75	5.1508	0.00342	4.73879	155,249

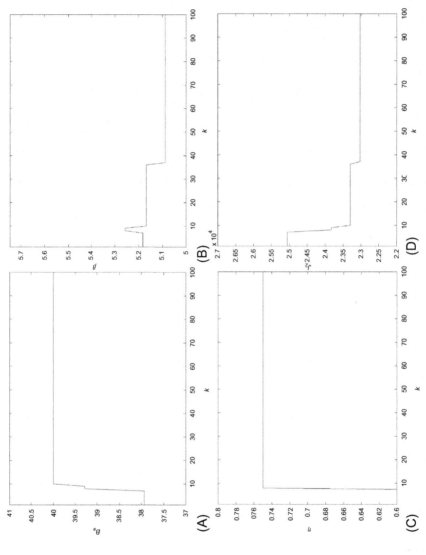

Fig. 4.5 Takagi-Sugeno PI-FC tuning parameters and objective function J_2 evolution vs iteration index in GWOPSO algorithm running: B_e vs k (A), β vs k (B), η vs k (C), and J_2 vs k (D).

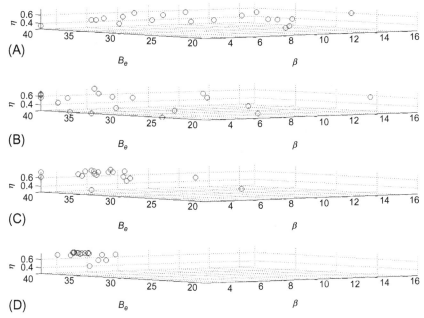

Fig. 4.6 Evolution of vector solution ρ to the optimization problem (1.80) solved by GWO PSO algorithm in the search domain D_ρ for four values of iteration index k: $k=1$ (A), $k=15$ (B), $k=60$ (C), and $k=100$ (D).

References

Chutani, S., Singh, J., 2018. Use of modified hybrid PSOGSA for optimum design of RC frame. J. Chin. Inst. Eng. 41 (4), 342–352.

David, R.-C., Precup, R.-E., Petriu, E.M., Purcaru, C., Preitl, S., 2012. PSO and GSA algorithms for fuzzy controller tuning with reduced process small time constant sensitivity. In: Proceedings of 2012 16th International Conference on System Theory, Control and Computing, Sinaia, Romania, pp. 1–6.

Dubey, H.M., Pandit, M., Panigrahi, B.K., Udgir, M., 2013. A novel swarm intelligence based gravitational search algorithm for combined economic and emission dispatch problems. In: Panigrahi, B.K., Suganthan, P.N., Das, S., Dash, S.S. (Eds.), Swarm, Evolutionary and Memetic Computing. In: Lecture Notes in Computer Science, vol. 8297. Springer-Verlag, Berlin, Heidelberg, pp. 568–579.

Jadidi, Z., Muthukkumarasamy, V., Sithirasenan, E., 2013. Metaheuristic algorithms based flow anomaly detector. In: Proceedings of 19th Asia-Pacific Conference on Communications, Bali, Indonesia, pp. 717–722.

Jiang, S., Ji, Z., Shen, Y., 2014. A novel hybrid particle swarm optimization and gravitational search algorithm for solving economic emission load dispatch problems with various practical constraints. Int. J. Electr. Power Energy Syst. 55, 628–644.

Jitkongchuen, D., 2015. A hybrid differential evolution with grey wolf optimizer for continuous global optimization. In: Proceedings of 7th International Conference on Information Technology and Electrical Engineering, Chiang Mai, Thailand, pp. 51–54.

Kennedy, J., Eberhart, R.C., 1995a. Particle swarm optimization. In: Proceedings of 1995 IEEE International Conference on Neural Networks, Perth, Australia, pp. 1942–1948.

Kennedy, J., Eberhart, R.C., 1995b. A new optimizer using particle swarm theory. In: Proceedings of 6th International Symposium on Micro Machine and Human Science, Nagoya, Japan, pp. 39–43.

Khadanga, R.K., Satapathy, J.K., 2015. Time delay approach for PSS and SSSC based coordinated controller design using hybrid PSO-GSA algorithm. Int. J. Electr. Power Energy Syst. 71, 262–273.

Lian, C., Zeng, Z., Yao, W., Tang, H., 2013. Displacement prediction of landslide based on PSOGSA-ELM with mixed kernel. In: Proceedings of 2013 Sixth International Conference on Advanced Computational Intelligence, Hangzhou, China, pp. 52–57.

Mirjalili, S., Hashim, S.Z.M., 2010. A new hybrid PSOGSA algorithm for function optimization. In: Proceedings of 2010 International Conference on Computer and Information Application, Tianjin, China, pp. 374–377.

Pei, J., Liu, X., Pardalos, P.M., Fan, W., Yang, S., Wang, L., 2014. Application of an effective modified gravitational search algorithm for the coordinated scheduling problem in a two-stage supply chain. Int. J. Adv. Manuf. Technol. 70 (1–4), 335–348.

Prajna, K., Reddy, K.V.V.S., Sasi Bhushan Rao, G., Uma Maheswari, R., 2015. A comparative study of BA, APSO, GSA, hybrid PSOGSA and SPSO in dual channel speech enhancement. Int. J. Speech Technol. 18 (4), 663–671.

Purcaru, C., Precup, R.-E., Iercan, D., Fedorovici, L.-O., David, R.-C., 2013. Hybrid PSO-GSA robot path planning algorithm in static environments with danger zones. In: Proceedings of 2013 17th International Conference on System Theory, Control and Computing, Sinaia, Romania, pp. 434–439.

Singh, N., Singh, S., Singh, S.B., 2017. A new hybrid MGBPSO-GSA variant for improving function optimization solution in search space. Evol. Bioinforma. 13, 1–13. https://doi.org/10.1177/1176934317699855.

Singh, N., Singh, S.B., 2017a. A novel hybrid GWO-SCA approach for optimization problems. Int. J. Eng. Sci. Technol. 20 (6), 1586–1601.

Singh, N., Singh, S.B., 2017b. Hybrid algorithm of particle swarm optimization and grey wolf optimizer for improving convergence performance. J. Appl. Math. 2017, 1–15. https://doi.org/10.1155/2017/2030489.

Tawhid, M.A., Ali, A.F., 2017. A hybrid grey wolf optimizer and genetic algorithm for minimizing potential energy function. Memetic Comput. 9 (4), 347–359.

Zhang, S., Luo, Q., Zhou, Y.-Q., 2017. Hybrid grey wolf optimizer using elite opposition-based learning strategy and simplex method. Int. J. Comput. Intell. Appl. 16 (2), 1–37. https://doi.org/10.1142/S1469026817500122.

CHAPTER 5

Conclusions

Contents

5.1 Performance comparison of nature-inspired optimization algorithms
in the optimal tuning of Takagi-Sugeno proportional-integral fuzzy controllers   115
5.2 A sample of experimental results   119
5.3 Extensions to proportional-integral-derivative fuzzy controllers   121
5.4 Extensions to type-2 fuzzy controllers   123
5.5 Extensions to tensor product-based model transformation controllers   124
5.6 Extensions to evolving fuzzy systems and controllers   127
5.7 Perspectives of nature-inspired algorithms applied to fuzzy control   131
References   134

Abstract

This chapter provides a systematic performance comparison of nature-inspired optimization algorithms in the optimal tuning of Takagi-Sugeno proportional-integral fuzzy controllers (FCs) for the position control of servo systems. Extensions to type-2 fuzzy control, tensor product-based model transformation controllers, and evolving fuzzy systems are discussed and related to optimal tuning of FCs. Perspectives of nature-inspired algorithms applied to fuzzy control are formulated.

Keywords: Evolving fuzzy systems, Performance comparison, Takagi-Sugeno proportional-integral-derivative fuzzy controllers, Tensor product-based model transformation controllers, Type-2 fuzzy control

5.1 Performance comparison of nature-inspired optimization algorithms in the optimal tuning of Takagi-Sugeno proportional-integral fuzzy controllers

If Chapters 2–4 were dedicated to the introduction of several nature-inspired algorithms and presenting the solutions they provided for the optimization problems defined in Chapter 1, this chapter gives a performance comparison of these algorithms with regard to the values of the objective functions in the optimization problems.

As previously mentioned, the algorithms required several restarts before the optimal values were obtained in order to overcome the random characteristic common to all nature-inspired algorithms. Therefore, the averages of the best values obtained for each combination of objective function and weighting parameter are taken into consideration. The best values for objective function are the smallest values in the context of the optimization problems defined in Eqs. (1.79)–(1.82) that target the minimization of several objective functions.

The first comparison criterion is represented by the average value of each objective function defined in Eqs. (1.75)–(1.78) obtained by running each nature-inspired optimization algorithm presented in previous chapters. The notation for this criterion is $Avg(J_{\Theta\min})$, and it is calculated in terms of

$$Avg\left(J_{\Theta}\min\right) = \frac{1}{N_{best}} \sum_{j=1}^{N_{best}} J_{\Theta}\min^{(j)}, \quad \Theta = 1, \ldots, 4, \qquad (5.1)$$

where $J_{\Theta\min}$ is the value of the objective function obtained by running a certain nature-inspired optimization algorithm considering one of the four objective function expressions given in Eqs. (1.75)–(1.78), N_{best} represents the number of best values (i.e., the smallest values) obtained for each objective function, and the superscript j, $j = 1, \ldots, N_{best}$, indicates the value of the objective function $J_{\Theta\min}$ obtained by one of the best N_{best} runs of a certain nature-inspired optimization algorithm, so $J_{\Theta\min}^{(j)}$ is the value of the objective function $J_{\Theta\min}$ obtained by the run j, $j = 1, \ldots, N_{best}$, of a certain nature-inspired optimization algorithm.

The results presented in Tables 2.1–2.4, 3.1–3.3, 4.1, and 4.2 represent the average of the best $N_{best} = 5$ values obtained for each objective function. However, despite obtaining different values, the same qualitative conclusions can be drawn for other values of N_{best}.

By analyzing the performance of all proposed nature-inspired algorithms it can be observed that no algorithm has a dominant position compared to the others, as the values of the proposed comparison criterion are relatively close. In addition, the best values for each objective function may be given by different algorithms in each of these cases. If the comparison is restricted to the adaptive and regular versions of the nature-inspired optimization algorithms, it can be observed that the adaptive version of each algorithm outperforms the regular version.

The second comparison criterion is based on the performance index referred to as convergence speed (c_s). As defined by Precup et al. (2013b), the convergence speed represents the number of evaluations of the objective functions until their minimum value is found.

This approach is of special importance for the nature-inspired optimization algorithms applied to the optimal tuning of the parameters of controllers. The algorithm complexity analysis is generally used in the analysis of numerical algorithms including optimization ones in the general context of computer science, where assessing the amount of required resources to execute these algorithms is discussed. The algorithm complexity analysis is not carried out in this book, since:

- The optimization algorithms treated in this book are designed to work with a fixed number of inputs (i.e., the variables of the objective functions, namely the tuning parameters of the controllers), however, the algorithms used in computer science are designed to work with inputs of arbitrary length, but the number of inputs set in the algorithms treated in this book is fixed to three in order to have a reasonable dimension of the search space.
- The optimization algorithms are executed offline and only the evaluation of the objective function, conducted by simulations and/or experiments, requires strong time resources on the control system side, which are much costlier compared to the resources on the algorithm execution side.

In this context, the convergence speed c_s is an indication on the complexity of these algorithms. However, in the general application of these algorithms to various optimization problems involving different objective functions with several numbers of variables, the algorithm complexity analysis becomes strictly necessary. The data corresponding to this second performance criterion represents the degree of algorithm iterations coverage before finding the final solution.

The results presented in Tables 5.1 and 5.2 contain the average values of the convergence speed c_s calculated for five best runs, used for the previous comparison criterion as well. Each algorithm was considered for each objective function defined in Eqs. (1.75)–(1.78).

Table 5.1 Average values of convergence speed c_s in the minimization of J_Θ, $\Theta = 1, \ldots, 4$, using PSO, GSA, CSS, GWO, adaptive GSA, and adaptive CSS algorithms

J_Θ	c_s PSO	c_s GSA	c_s CSS	c_s GWO	c_s adaptive GSA	c_s adaptive CSS
J_1	218	1554	757	202	1298	1306
J_2	1906	1399	1358	2005	1718	1905
J_3	1647	596	1596	1682	1768	828
J_4	1618	499	1325	1587	1480	1659

Table 5.2 Average values of convergence speed c_s in the minimization of J_Θ, $\Theta = 1, \ldots, 4$, using fuzzy logic-based adaptive GSA, hybrid PSOGSA and hybrid GWOPSO algorithms

J_Θ	c_s fuzzy logic-based adaptive GSA	c_s hybrid PSOGSA	c_s hybrid GWOPSO
J_1	1203	190	178
J_2	1662	1660	1843
J_3	1680	723	1541
J_4	1395	1826	1523

Tables 5.1 and 5.2 show that the results obtained when the adaptive GSA and CSS algorithms were used are consistently superior to those offered by the nonadaptive algorithm versions. Local minima traps are also avoided when using the adaptive algorithm versions by the introduction of improved exploration and exploitation capabilities. On the negative side, finding the solution after a predefined number of iterations may result in longer time runs for the adaptive GSA and CSS algorithms.

As the performance index c_s focuses on how fast a solution is found, it can miss other relevant information on the overall solution's quality. This limitation can be mitigated by the introduction of the third performance index, namely the accuracy rate a_r defined as the percent standard deviation of the objective functions obtained by running a certain optimization algorithm divided by the average value $Avg(J_{\Theta min})$ of each objective function given in Eqs. (1.75)–(1.78) obtained by running a certain nature-inspired optimization algorithm:

$$a_r = StDev^{\%}(J_{\Theta min}) = 100\,\frac{StDev(J_{\Theta min})}{Avg(J_{\Theta min})}, \quad \Theta = 1, \ldots, 4, \qquad (5.2)$$

where the average value $Avg(J_{\Theta min})$ of each objective function given in Eqs. (1.75)–(1.78) obtained by running a certain nature-inspired optimization algorithm is defined in Eq. (5.1), the standard deviation $StDev(J_{\Theta min})$ is calculated in terms of

$$StDev(J_{\Theta min}) = \sqrt{\frac{1}{N_{best} - 1}\sum_{j=1}^{N_{best}} \left(J_{\Theta min}^{(j)} - Avg(J_{\Theta min})\right)^2}, \quad \Theta = 1, \ldots, 4,$$

$$(5.3)$$

and the other notations are explained in relation with Eq. (5.1).

Tables 5.3 and 5.4 illustrate the average values of a_r based on the best $N_{best} = 5$ runs that are in the case of previous comparison indices as well.

Table 5.3 Average values of accuracy rate a_r in the minimization of J_Θ, $\Theta = 1, \ldots, 4$, using PSO, GSA, CSS, GWO, adaptive GSA and adaptive CSS algorithms

J_Θ	a_r PSO	a_r GSA	a_r CSS	a_r GWO	a_r Adaptive GSA	a_r Adaptive CSS
J_1	0	0.0044	0.1887	0.1035	0.00015	0.2131
J_2	0.2702	2.1778	0.5706	2.3894	0.6448	0.0012
J_3	0.1947	0.2775	15.7606	0.3646	2.5354	2.9227
J_4	0.0036	4.7067	22.7008	4.9811	1.5897	0.2075

Table 5.4 Average values of accuracy rate a_r in the minimization of J_Θ, $\Theta = 1, \ldots, 4$, using fuzzy logic-based adaptive GSA, hybrid PSOGSA and hybrid GWOPSO algorithms

J_Θ	a_r fuzzy logic-based adaptive GSA	a_r hybrid PSOGSA	a_r hybrid GWOPSO
J_1	0.2015	0.2071	0.2218
J_2	0.0011	5.6485	6.1783
J_3	2.5861	7.6804	8.4913
J_4	0.1974	8.6845	9.5143

These values correspond to the values of a_r presented comparatively for all nature-inspired algorithms discussed in this book and each objective function.

The results outline that PSO and adaptive GSA and CSS algorithms have an improved search process which leads to the convergence to the optimal solution at the end of the search process. The accuracy rate values clearly show that the adaptive version of the algorithms have a higher accuracy rate of finding closer optimal solutions, thus increasing the confidence in the solutions provided. Therefore, all search iterations are used compared to the non–adaptive GSA and CSS algorithms, which converge too early, leading to an increased chance of getting trapped in local minima situations and an unnecessary computational cost.

5.2 A sample of experimental results

The solutions obtained from the implementation of the nature-inspired optimization algorithms to solve the optimization problems defined in Eqs. (1.79)–(1.82) were tested on the experimental setup described in Chapter 1. Several experimental results have been reported by David et al. (2013), Precup et al. (2012a, b, 2013a, b, 2014a, 2017a, b), and Roman et al. (2018) and validate the solution offered by means of nature-inspired optimization algorithms, the design approach, and the fuzzy controllers (FCs).

As used in the evaluations of the objective functions presented in Chapters 2–4, the dynamic regimes considered were characterized by the $r = r_0 = 40$ rad step-type modification of the reference input and zero disturbance input, $d = 0$. An example of real-time experimental results of the fuzzy control system with the Takagi-Sugeno PI-FC (with the optimal parameters obtained by the PSO-based minimization of the objective function J_2) and the PI controller (tuned for the same optimal parameters in the linear part of the Takagi-Sugeno PI-FC) is presented in Fig. 5.1.

The results of the real-time experiments present the output of the system and of the controller and prove the disturbance rejection and the presence of the insensitivity zone in the real-world controlled process (CP). An insensitivity zone of such magnitude in the actuator is therefore difficult to use for precise positioning because the control signal is subjected to oscillations. Fig. 5.1 shows, as expected, the performance improvement exhibited by the control system with the Takagi-Sugeno PI-FC compared to the control system with the linear (PI) controller, namely the settling time is reduced.

Since the fuzzy control systems with controllers optimally tuned using other nature-inspired algorithms exhibit behaviors similar to those presented in

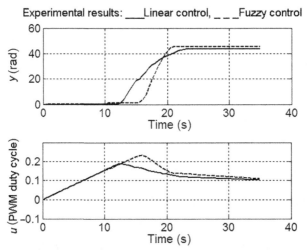

Fig. 5.1 Real-time experimental results expressed as controlled output and control signal of the control system with the PI controller *(dashed line)* and of the control system with the Takagi-Sugeno PI-FC *(solid line)*. *(From Precup, R.-E., David, R.-C., Petriu, E.M., Radac, M.-B., Preitl, S., Fodor, J., 2013. Evolutionary optimization-based tuning of low-cost fuzzy controllers for servo systems. Knowl.-Based Syst. 38, 74–84).*

Fig. 5.1, other system responses are not given. This is justified by the fact that all algorithms lead to practically the same solution to the optimization problems defined in Eqs. (1.79)–(1.82). However, as shown in the previous chapters of this book, the values of the tuning parameters of the FCs are different.

5.3 Extensions to proportional-integral-derivative fuzzy controllers

The proportional-derivative fuzzy controllers (PD-FCs) are designed using the approach described in Chapter 1 and dedicated to PI-FCs. The design is carried out in terms of a PD-FC structure obtained from the PI-FC with output integration with the structure shown in Fig. 1.9 by dropping out the output integrator.

The PD-FCs are used, as PI-FCs, to control the servo system considered in this book. The PD-FCs are also recommended for servo system control if the process models of such servo systems contain one large time constant to be canceled by the controllers. The PI-FCs can also be used to control the servo system models described in Chapter 1 with an additional large time constant, but the ESO method cannot be applied in the linear case to tune the PI controllers. The difference between the PI-FCs and the PD-FCs in controlling the servo system models considered in this book is that the zero steady-state control error can be ensured by the control systems with PI-FCs for any constant disturbance inputs, whereas the control systems with PD-FCs cannot ensure the zero steady-state control error as constant disturbances are applied to the process input (control signal).

The Takagi-Sugeno proportional-integral-derivative fuzzy controllers (PID-FCs) are generally recommended to control servo systems with process models that contain two large time constants, which can be canceled by the controllers. The combination of ESO method and cancellation makes the PID-FCs able to control the servo system models described in Chapter 1 with an additional large time constant. Although in the linear case PID controllers are not recommended in situations where strong variations of the reference and/or disturbances occur due to the zeros specific to the controllers, the PID-FCs can cope with such variations because of the limitations specific to the FC structure.

The PID-FC structure is shown in Fig. 5.2A, where the TISO-FC^{PD} block and the TISO-FC^{PI} block are identical to the TISO-FC block shown in Fig. 1.9, u_k^{PD} is the control signal of the PD controller, and Δu_k^{PI} is the increment of control signal calculated by the PI controller. The fuzzification

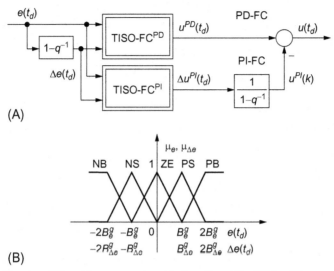

(A)

(B)

Fig. 5.2 Structure (A) and input membership functions (B) of Takagi-Sugeno PID-FC.

is carried out by means of the input membership functions shown in Fig. 5.2B, where $g \in \{PI, PD\}$. Fig. 5.2A shows that the PID–FC structure consists of a parallel connection of PD-FC and PI-FC.

The design of the PID–FC is carried out using the same design approach described in Chapter 1. However, the equivalence to the linear PID controller has to be ensured in terms of the manipulation of the input–output characterization of the linear PID controller, which is transformed into the nonlinear parallel structure shown in Fig. 4.2. The transfer function $C(s)$ of the linear PID controller is

$$C(s) = k_C \left(1 + \frac{1}{sT_i} + sT_d \right) = k_c \frac{(1 + sT_{c1})(1 + sT_{c2})}{s},$$

$$k_c = \frac{k_C}{T_i}, \quad T_{c1,2} = \frac{T_i \left(1 \pm \sqrt{1 - 4T_d/T_i} \right)}{2},$$

(5.4)

where T_d is the derivative time constant, and T_{c1} and T_{c2} are the controller time constants.

Setting the sampling period T_s, Tustin's method leads to the discrete-time PID controller with the transfer function $C(q^{-1})$.

$$C(q^{-1}) = \frac{\rho_0 + \rho_1 q^{-1} + \rho_2 q^{-2}}{1 - q^{-1}},$$

$$\rho_0 = k_c \left(1 + \frac{T_d}{T_s} \right), \quad \rho_1 = -k_c \left(1 + \frac{2T_d}{T_s} - \frac{T_s}{T_i} \right), \quad \rho_2 = k_c \frac{T_d}{T_s}.$$

(5.5)

The relationship (5.5) is then expressed as parallel connection of a discrete-time PD controller and a discrete-time PI controller with the superscripts PD and PI, respectively, associated to Fig. 5.2:

$$C(q^{-1}) = K_P^{PD}(1 - q^{-1} + \mu^{PD}) - K_P^{PI} \frac{1 - q^{-1} + \mu^{PI}}{1 - q^{-1}},$$

$$K_P^{PD} = \rho_2, \quad \mu^{PD} = -\frac{\rho_1}{\rho_2}, \quad K_P^{PI} = \rho_2 - \rho_1 - \rho_0, \quad \mu^{PI} = \frac{2\rho_2}{\rho_2 - \rho_1 - \rho_0}. \tag{5.6}$$

Therefore, it is justified to carry out the separate design of the PI-FC and the PD-FC illustrated in Fig. 5.2 using the same design approach. However, more parameters should be optimally tuned by nature-inspired algorithms in steps 3 and 4.

The PI-FC can be involved in two-degrees-of-freedom (2-DOF) FCs by the fuzzification of some of the controller blocks in 2-DOF linear controllers in order to improve the control system performance. The 2-DOF PI-FCs exhibit performance improvements over 1-DOF FCs. Similar controller structures can be formulated as state-feedback control systems as well.

5.4 Extensions to type-2 fuzzy controllers

The extensions of FCs from type-1 FCs treated in this book to type-2 FCs prove additional performance improvement as reported by Castillo et al. (2011) focusing on control problems with a high degree of uncertainty.

The structure of a type-2 fuzzy control system is presented in Fig. 5.3, which is a generalized version of the type-1 fuzzy control system structure presented in Fig. 1.1. The type-2 fuzzy control system is considered as a single input system with respect to the reference input r and as a single output system with respect to the controlled output y. The second input applied to the CP is the disturbance input d.

Fig. 5.3 illustrates the operation principle of a type-2 FC, with the following variables and modules: (1) the crisp inputs, (2) the fuzzification module, (3) the fuzzified inputs, (4) the inference module, (5) the type-2 fuzzy conclusions, (6) the type reducer module, (7) the type-1 fuzzy conclusions, (8) the defuzzification module, and (9) the crisp output. The rest of the variables in Fig. 5.3 are defined in relation to Fig. 1.1.

The design and tuning of type-2 FCs is mainly based on nature-inspired optimization algorithms. Some representative applications of type-2 FCs have been discussed by Linda and Manic (2011), Oh et al. (2011), Castillo and Melin (2012, 2014), Mohammadzadeh et al. (2014), Precup et al.

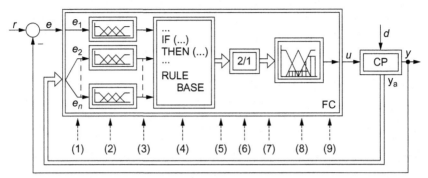

Fig. 5.3 Type-2 fuzzy control system structure. *(From Precup, R.-E., Angelov, P., Costa, B.S.J., Sayed-Mouchaweh, M., 2015. An overview on fault diagnosis and nature-inspired optimal control of industrial process applications. Comput. Ind. 74, 75–94).*

(2015a), Wang et al. (2015), Kayacan and Khanesar (2016), Baghbani et al. (2018), and Hong and Nguyen (2018).

5.5 Extensions to tensor product-based model transformation controllers

The tensor product (TP)-based model transformation is capable of transforming a dynamic system model, given over a bounded domain, into the TP model form, including polytopic and Takagi-Sugeno fuzzy model forms. The TP-based model transformation may be defined as one numerical method capable of transforming the linear parameter-varying (LPV) dynamic models into parameter-varying weighted combinations of parameter independent (constant) system models under the form of linear time-invariant (LTI) systems. As shown by Baranyi (2016), this transformation of LPV models is uniform in both theoretical and algorithmic execution and considers different optimization constraints. The main advantage of TP model transformation in modifying the given LPV models to varying convex combinations of LTI models is that the linear matrix inequality (LMI)-based control design frameworks can be applied immediately to the resulting affine models in order to get a tractable and improved performance of the FCSs.

As pointed out by Precup and Hellendoorn (2011), the popular transfer function of the product decision operator-based Takagi-Sugeno fuzzy models and the function of the TP models are the same from the analytical point of view in general cases. The main operational difference between them is that the Takagi-Sugeno fuzzy model originally means a fuzzy combination of locally linearized LTI models, where the locality is expressed by the shape of the input membership functions. However, in the case of TP model the weighting functions (which correspond to the membership functions in the fuzzy models) may not have locality, they spread in the whole interval of interest, so the LTI components of the model cannot readily be assigned to a definite operation point. They are mostly vertexes of a polytopic structure. Therefore, the Takagi-Sugeno fuzzy model originally is a fuzzy combination of linearized operation points (LTI systems are close to local models), while the TP model is originally a polytopic structure (LTI systems are the vertex models of a convex hull of the model, they may be relatively far from any linearized operation points). In other words, an LTI system affects a fuzzy local area in the case of Takagi-Sugeno models, whereas in the case of TP models an LTI system affects the whole operation domain, but according to the weighting functions. Since these ideas are applied nowadays in combination in both Takagi-Sugeno and TP models, the difference is not important and justifies the presentation in the sequel of some details on TP models although this book is dedicated to fuzzy systems and control.

The TP-based model transformation generates two kinds of polytopic models. Initially, it reconstructs the high-order singular value decomposition (HOSVD)-based canonical form of the LPV models. The major advantage of HOSVD is its ability to decompose a given N-dimensional tensor into a full orthonormal system in a special ordering of higher-order singular values.

Regarding the variety of well acknowledged and implemented identification techniques, it is difficult to derive the uniform representation of the designed LPV model forms and the forms resulted from the identification. Hence the TP-based model transformation represents a viable solution to that situation, which leads to the link between the model transformation and the LMIs.

Another advantage of TP-based model transformation is that it allows for the modification of the parameter varying convex combination according to designer's option. The type of the convex combination considerably influences the further LMI design and resulting control performance. Therefore, the design can be based on the manipulation of the convex hull besides the manipulation of the LMIs.

Making use of the TP-based model transformation, different optimization and convexity constraints can be considered and the transformations can be executed as well without any analytical interactions within less time.

Thus, the transformation replaces the usual analytical conversions. These aspects make the TP models and controllers subject to be optimally tuned using nature-inspired algorithms.

The TP-based model transformation is also used, by generalization, to obtain polytopic quasi-linear parameter-varying (qLPV) models as pointed in the following. Accepting an N-dimensional bounded parameter vector $\mathbf{p}(t)$, let the qLPV model be

$$
\begin{aligned}
\dot{\mathbf{x}}(t) &= \mathbf{A}(\mathbf{p}(t))\mathbf{x}(t) + \mathbf{B}(\mathbf{p}(t))\mathbf{u}(t), \\
\mathbf{y}(t) &= \mathbf{C}(\mathbf{p}(t))\mathbf{x}(t) + \mathbf{D}(\mathbf{p}(t))\mathbf{u}(t),
\end{aligned}
\tag{5.7}
$$

with the input (control signal) vector $\mathbf{u}(t) \in \mathfrak{R}^m$, output vector $\mathbf{y}(t) \in \mathfrak{R}^l$, and state vector $\mathbf{x}(t) \in \mathfrak{R}^q$. The parameter vector fulfills the condition.

$$
\mathbf{p}(t) \in \Omega, \ \Omega = [a_1, b_1] \times [a_2, b_2] \times \cdots \times [a_N, b_N] \subset \mathfrak{R}^N,
\tag{5.8}
$$

where Ω is a closed hypercube. The system matrix $\mathbf{S}(\mathbf{p}(t))$.

$$
\mathbf{S}(\mathbf{p}(t)) = \begin{pmatrix} \mathbf{A}(\mathbf{p}(t)) & \mathbf{B}(\mathbf{p}(t)) \\ \mathbf{C}(\mathbf{p}(t)) & \mathbf{D}(\mathbf{p}(t)) \end{pmatrix} \in \mathfrak{R}^{(l+q)\times(m+q)}
\tag{5.9}
$$

is a parameter-varying object.

For any parameter vector $\mathbf{p}(t)$ the system matrix $\mathbf{S}(\mathbf{p}(t))$ in Eq. (5.9) is next expressed as

$$
\mathbf{S}(\mathbf{p}(t)) = \mathbf{S} \overset{N}{\underset{n=1}{\otimes}} \mathbf{w}(p_n(t)),
\tag{5.10}
$$

where $\mathbf{S}$ is the matrix of LTI vertex systems, $\mathbf{w}(p_n(t))$ are the weighting function matrices, m is the number of system inputs, and q is the number of state variables. In other words, the convex state-space TP model characterized by Eq. (5.10) describes a qLPV state-space model for any parameter vector $\mathbf{p}(t)$ as the convex combination of LTI system matrices.

The (finite element) TP model defined in Eq. (5.10) is convex only if the weighting functions fulfill the condition

$$
\begin{aligned}
w_{n,i}(\mathbf{p}_n(t)) &\in [0, 1], \ \ \forall n, i, \mathbf{p}_n(t), \\
\sum_{i=1}^{I_n} w_{n,i}(\mathbf{p}_n(t)) &= 1, \ \ \forall n, \mathbf{p}_n(t).
\end{aligned}
\tag{5.11}
$$

The LMI-based controller design methods can immediately be applied after the transformation of the qLPV model Eq. (5.7) given in HOSVD-based canonical form to the TP model form expressed in Eqs. (5.10) and (5.11). In other words, the TP-based model transformation is to be used and executed before utilizing the LMI design, that is, when the LMI design is started the global weighting functions are already defined. This mathematical characterization is also used to treat the TP models in fuzzy model manner, thus justifying once more the inclusion of their brief description in this book.

The TP-based model transformation is actually implemented as TP controllers, which are actually nonlinear state feedback controllers. The TP controllers are currently used in many important theoretical topics and nonlinear control applications that include sliding surface design (Korondi, 2006), aeroelastic wing section (Baranyi and Takarics, 2014; Szöllősi and Baranyi, 2018), analysis of design of systems modeled by qLPV models (Baranyi, 2015), multitank systems (Precup et al., 2015b), interpolation and rule reduction (Campos et al., 2015), tumor growth control (Kovács and Eigner, 2018), spacecraft systems (Gong et al., 2018), and magnetic levitation systems (Hedrea et al., 2017, 2018).

5.6 Extensions to evolving fuzzy systems and controllers

As pointed out by Precup et al. (2018), the concept of evolving fuzzy (rule-based) controllers was coined by P. Angelov in 2001 and was further developed in his later works (Angelov et al., 2001; Angelov, 2002; Angelov and Buswell, 2002; Angelov and Filev, 2002, 2003; Sadeghi-Tehran et al., 2012). These controllers are related to evolving Takagi-Sugeno fuzzy models, for which the rule bases are computed by a learning process that comprises continuous online rule base learning as shown in the recent papers (Blažič et al., 2014; Precup et al., 2014b; Leite et al., 2015; Za'in et al., 2017; Andonovski et al., 2018; Lughofer et al., 2018).

Using the details provided by Precup et al. (2014b), some implementation aspects concerning an incremental online identification algorithm are given. The flowchart of the incremental online identification algorithm is presented in Fig. 5.4, and steps S1 to S7 of this algorithm are briefly described as follows.

S1. The rule base structure is initialized. All parameters in the rule antecedents are set to obtain one rule, $n_R = 1$ (n_R–the number of rules). The parameters of the evolving Takagi-Sugeno fuzzy models are calculated by subtractive clustering for the first data point $\mathbf{p}_1$. The notation $\mathbf{p}_k$ is used for the data point $\mathbf{p}$ at the discrete-time step k (that is also the index of the current sample), which belongs to the input-output data set $\{\mathbf{p}_k | k = 1, \dots, D\} \subset \mathfrak{R}^{n+1}$.

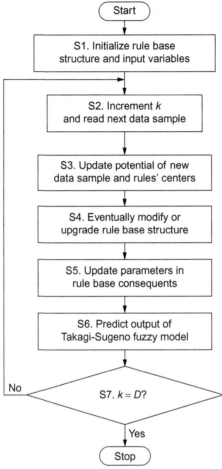

Fig. 5.4 Flowchart of incremental online identification algorithm. *(From Precup, R.-E., Filip, H.-I., Radac, M.-B., Petriu, E.M., Preitl, S., Dragos, C.-A., 2014b. Online identification of evolving Takagi-Sugeno-Kang fuzzy models for crane systems. Appl. Soft Comput. 24, 1155–1163).*

$$\mathbf{p}_k = \left[p_k^1 \; p_k^2 \cdots p_k^{n+1} \right]^T, \quad \mathbf{p} = \left[\mathbf{z}^T \; y \right]^T = \left[z_1 \; z_2 \cdots z_n \; y_l \right]^T$$
$$= \left[p^1 \; p^2 \cdots p^n \; p^{n+1} \right]^T \in \mathfrak{R}^{n+1}, \tag{5.12}$$

where D is the number of input–output data points or data samples, and $\mathbf{z}$ is the input vector.

The rule base of Takagi–Sugeno fuzzy models is expressed as follows making use of affine rule consequents:

Rule i: IF z_1 IS LT_{i1} AND ... AND z_n IS LT_{in} THEN

$$y_l^i = a_{i0} + a_{i1} z_1 + \cdots + a_{in} z_n, i = 1, \dots, n_R, \tag{5.13}$$

where z_j, $j = 1, \dots, n$, are the input (or scheduling) variables, n is the number of input variables, $LT_{i\,j}$, $i = 1, \dots, n_R, j = 1, \dots, n$, are the input linguistic terms, y_l^i is the output of the local model in the rule consequent of rule i, $i = 1, \dots, n_R$, and $a_{i\,\chi}$, $i = 1, \dots, n_R, \chi = 0, \dots, n$, are the parameters in the rule consequents.

Considering the algebraic product t-norm as an AND operator and the weighted average defuzzification method, the expression of the Takagi-Sugeno fuzzy model output y_l is

$$
y_l = \frac{\displaystyle\sum_{i=1}^{n_R} \tau_i y_l^i}{\displaystyle\sum_{i=1}^{n_R} \tau_i} = \sum_{i=1}^{n_R} \lambda_i y_l^i, \quad y_l^i = \begin{bmatrix} 1 & \mathbf{z}^T \end{bmatrix} \boldsymbol{\pi}_i,
$$

$$
\lambda_i = \tau_i / \left[\sum_{i=1}^{n_R} \tau_i \right], \quad i = 1, \dots, n_R,
$$

(5.14)

where the $\tau_i(\mathbf{z})$ is firing degree of rule i and $\lambda_i(\mathbf{z})$ the normalized firing degree, and the parameter vector of rule i is $\boldsymbol{\pi}_i$, $i = 1, \dots, n_R$, $\tau_i(\mathbf{z})$ is calculated in terms of

$$
\begin{aligned}
\tau_i(\mathbf{z}) &= \mathrm{AND}(\mu_{i1}(z_1), \mu_{i2}(z_2), \dots, \mu_{in}(z_n)) \\
&= \mu_{i1}(z_1) \cdot \mu_{i2}(z_2) \cdot \dots \cdot \mu_{in}(z_n), \quad i = 1, \dots, n_R,
\end{aligned}
$$

(5.15)

and the parameter vector of rule i is $\boldsymbol{\pi}_i$

$$
\boldsymbol{\pi}_i = \begin{bmatrix} a_{i0} & a_{i1} & a_{i2} & \cdots & a_{in} \end{bmatrix}^T, \quad i = 1, \dots, n_R.
$$

(5.16)

The main parameters specific to the incremental online identification algorithm are initialized according to (Angelov and Filev, 2004).

$$
\begin{aligned}
\hat{\boldsymbol{\theta}}_1 &= \left[\left(\boldsymbol{\pi}_1^T \right)_1 \ \left(\boldsymbol{\pi}_2^T \right)_1 \cdots \left(\boldsymbol{\pi}_{n_R}^T \right)_1 \right]^T = \begin{bmatrix} 0 & 0 \cdots 0 \end{bmatrix}^T, \quad \mathbf{C}_1 = \boldsymbol{\Omega}\mathbf{I}, \quad r_s = 0.4, \quad k \\
&= 1, \quad n_R = 1, \quad \mathbf{z}_1^* = \mathbf{z}_k, \quad P_1(\mathbf{p}_1^*) = 1,
\end{aligned}
$$

(5.17)

where $\mathbf{C}_k \in \mathfrak{R}^{n_R(n+1) \times n_R(n+1)}$ is the fuzzy covariance matrix (clusters), $\mathbf{I}$ is the $n_R(n+1)$th order identity matrix, $\boldsymbol{\Omega} = \mathrm{const}, \Omega > 0$, is a large number, $\hat{\boldsymbol{\theta}}_k$ is an estimation of the parameter vector in the rule consequents at time k, and r_s, $r_s > 0$, is the spread of all Gaussian input membership functions $\mu_{i\,j}$, $i = 1, \dots, n_R, j = 1, \dots, n$, of the fuzzy sets afferent to the input linguistic terms $LT_{i\,j}$.

$$
\mu_{ij}(z_j) = e^{-\dfrac{4\left(z_j - z_{ij}^*\right)^2}{r_s^2}}, \quad i = 1, \dots, n_R, \quad j = 1, \dots, n,
$$

(5.18)

where z_{ij}^* $i=1\ldots n_R, j=1\ldots n$, are the membership function centers, $\mathbf{p}_1^*$ in (8) is the first cluster center, $\mathbf{z}_1^*$ is the center of rule 1, equal to the projection of $\mathbf{p}_1^*$ on $\mathbf{z}$ axis in terms of (2), and $P_1(\mathbf{p}_1^*)$ is the potential of $\mathbf{p}_1^*$.

S2. The data sample index k is incremented, and the next data sample $\mathbf{p}_k$ that belongs to the input–output data set $\{\mathbf{p}_k \mid k=1,\ldots,D\} \subset \mathfrak{R}^{n+1}$ is collected.

S3. The potential of each new data sample $P_k(\mathbf{p}_k)$ and the potentials of the centers $P_k(\mathbf{p}_\eta^*)$ of existing rules (clusters) with the index η are recursively updated.

S4. If certain conditions given by Angelov and Filev (2004) are fulfilled, the rule base structure is modified or upgraded using potential of the new data in comparison with that of the existing rules' centers.

S5. The parameters in the rule consequents are updated using either the recursive least squares algorithm with a global objective function or the weighted recursive least squares algorithm with a locally weighted objective function. These updates allow for the computation of the updated $\hat{\boldsymbol{\theta}}_k$ and $\mathbf{C}_k$, $k=2,\ldots,D$.

S6. The output of the evolving TSK fuzzy model at the next discrete-time step $k+1$ is predicted as

$$\hat{y}_{k+1} = \boldsymbol{\psi}_k^T \hat{\boldsymbol{\theta}}_k, \tag{5.19}$$

with the notations

$$y = \boldsymbol{\psi}^T \boldsymbol{\theta}, \; \boldsymbol{\theta} = \begin{bmatrix} \boldsymbol{\pi}_1^T & \boldsymbol{\pi}_2^T & \cdots & \boldsymbol{\pi}_{n_R}^T \end{bmatrix}^T,$$
$$\boldsymbol{\psi}^T = \begin{bmatrix} \lambda_1[1 \; \mathbf{z}^T] & \lambda_2[1 \; \mathbf{z}^T] & \cdots & \lambda_{n_R}[1 \; \mathbf{z}^T] \end{bmatrix}. \tag{5.20}$$

S7. The algorithm continues with S2 until all data points of the input-output data set $\{\mathbf{p}_k \mid k=1,\ldots,D\}$ are collected.

The evolving FC structure suggested by Angelov (2004) is based on two ideas:

- indirect learning mechanism.
- evolving fuzzy rules.

The indirect learning mechanism is based on the approximation of the inverse dynamics of the process. The control scheme is considered in the framework of model-free control, and instead of feeding back the error between the process output and the reference input it feeds back the one-step delayed output. The incremental online identification algorithm described in this section is next applied. At each discrete-time step, the controller performs consecutive time steps of learning and control. In this way, the evolving FC learns and evolves its structure and its parameters online in the process of control itself. Some representative applications of such controllers that also include different problem settings are given by Zdešar et al. (2014), Andonovski et al. (2016), and Rong et al. (2017). As shown by Precup et al. (2016), nature-inspired optimization can be applied in combination with evolving fuzzy systems modeling.

5.7 Perspectives of nature-inspired algorithms applied to fuzzy control

This book provides a concise discussion on fuzzy control systems and the optimal tuning of FCs using nature-inspired algorithms. The focus is on Takagi-Sugeno PI-fuzzy control for position control of servo systems but since the presentation is transparent, the discussion and results can be applied relatively easily to various industrial applications with different FC structures.

Through a selective list of references, this book presents the historical motivation and the current research progress of Takagi-Sugeno FCs optimally tuned by current nature-inspired algorithms. As highlighted by Nguyen et al. (2019), great advances on both fundamental and application aspects of fuzzy control have been made in the last two decades, with a huge number of publications available on the topic. However, many interesting and important issues still remain a challenge, and they represent opportunities for fuzzy control systems design and tuning in the future. Starting with the results presented in the previous chapters, this book can be extended in the future directions briefly discussed as follows.

The presented solutions can be extended to other optimization problems that involve objective functions considered in the frequency domain. Robust stability and sensitivity can be considered in this regard.

The nature-inspired algorithms presented in this book can be also applied to fuzzy modeling of nonlinear dynamic processes by the parametric optimization using nature-inspired algorithms. This involved the optimal tuning of the parameters of fuzzy models included in several modules of the fuzzy model structure: fuzzification, rule base, inference engine, and/or defuzzification. The results regarding the fuzzification given by Precup et al. (2015c) can represent a starting point in this direction.

The approaches to design control systems with both Mamdani and Takagi-Sugeno FCs can be extended by including stability conditions for fuzzy control systems. These stability conditions represent constraints and should be inserted in appropriately defined optimization problems solved by nature-inspired algorithms.

As considered by Nguyen et al. (2019), the design of Takagi-Sugeno FCs and observers that do not share the same membership functions or premise variables with the Takagi-Sugeno fuzzy models of the process is another important topic. This can increase the flexibility of the controller and observer design in many cases (Nguang and Shi, 2003).

We consider that the extensions of fuzzy control to type-2 fuzzy systems, TP-based model transformation, and evolving fuzzy systems briefly treated in the previous sections are important topics in the field of fuzzy modeling and control. Their combination with nature-inspired optimization algorithms is a version to treat these topics in order to bridge the current gap between the theoretical developments and the real-world applications.

As far as the nature-inspired optimization algorithms are concerned, we believe that introducing and extending fuzzy logic in the tuning of these algorithms in order to limit the degrees of freedom represented by the free parameters and to improve their performance is a challenging topic. In addition, developing new solutions based on algorithm hybridization for alleviating the drawbacks encountered in standard versions should also be treated.

As outlined by Nguyen et al. (2019), in any control system, fault detection, diagnosis, and recovery are decisive to have a robust and resilient system operation. In this context, fuzzy control reconfiguration plays a crucial role to obtain a fault-tolerant control that can effectively handle severe actuator and/or sensor faults while still guaranteeing a desired closed-loop fuzzy control system performance. The combination of fault diagnosis discussed by Precup et al. (2015a) and nature-inspired optimization algorithms is one of the perspectives suggested in this context. On the one hand, the nature-inspired optimization algorithms can be applied to the optimal tuning of the models involved in fault diagnosis. On the other hand, the fault diagnosis methods can lead to conditions that can be embedded as additional constraints to the stability ones in the design and implementation of optimal controllers.

For a given nonlinear system, there is a problem to obtain as accurate as possible fuzzy model in terms of ensuring both accurate modeling and effective stability and performance analysis, and reducing the numerical complexity of controller design including stability analysis. The results provided by Zeng and Singh (1996) offer interesting ideas related to this open issue.

The results presented in this book in the context of fuzzy control are model-based because such approaches enable systematic frameworks for stability analysis and controller design of nonlinear systems in terms of optimal tuning of FCs. Moreover, the model-based approach to controller tuning is needed in this book due to the large number of evaluations of the objective functions. However, since there is always a problem finding a better and better accuracy of fuzzy models, the authors of this book think that model-free approaches to fuzzy control can be a strong perspective. The interest in developing controllers based on the input-output data has increased in the past two decades due the fact that sometimes for the practitioner it is very hard or impossible to obtain an accurate mathematical model of the process especially when the process is very complex and has strong nonlinearities. The controllers that use only the I-O (input-output) data from the process in their tuning, and actually carry out the experiment-based performance improvement, are built around data-driven techniques and algorithms; the most popular ones are iterative feedback tuning (Hjalmarsson et al., 1998; Preitl et al., 2006; Li et al., 2019; Son and Choi, 2018), model-free control (Fliess, 2013; Precup et al., 2017c; Bara et al., 2018; Safaei and Mahyuddin, 2018), model-free adaptive control (Zhu et al., 2017; Yu et al., 2018), active disturbance rejection control (Gao, 2006; Herbst, 2013), and virtual reference feedback tuning (Campi et al., 2002; Formentin et al., 2013).

The combination of data-driven model-free control with fuzzy control has been carried out by Precup et al. (2008a, b), however, this is an indirect approach that deals with first applying model-free tuning to the linear PI controllers and then incorporating the knowledge on these controllers in the tuning of PI-FCs. The first steps toward the direct application of data-driven model-free tuning to FCs have been done by Roman et al. (2017, 2018). Authors' opinion is that this approach represents a strong way to ensure the simple design of FCs focusing on performance improvement. In addition, the combination with nature-inspired optimization algorithms can lead to additional performance improvement, but the number of evaluations of the objective functions should be as small as possible. In addition, fractional-order control can also be involved in FCs as it leads to increased flexibility but associated with an increase in the dimension of the search space.

Another perspective of nature-inspired optimization algorithms, identified by Precup et al. (2015a), is the presentation of rather real-time experiments instead of digital simulation results. This is of exquisite importance in the context of optimal tuning of fuzzy control because of it being the only way to increase the interest in this nonlinear control strategy. The popularity of optimally tuned FCs will increase only if future applications such as, for example, those reported by Haber and Alique (2004), Filip (2008), Vaščák and Rutrich (2008), Precup et al. (2013c 2013), Purcaru et al. (2013), Osaba et al. (2016), Radiša et al. (2017) Sirbu and Dumitrache (2017), Alvarez Gil et al. (2018) Groumpos (2018), Guerrero et al. (2018), Vrkalović et al. (2018), will exhibit significantly better performance compared to existing ones. Thus, both researchers and practitioners will be attracted by this subject that will represent a guarantee for future successful industrial applications.

References

Alvarez Gil, R.P., Johanyák, Z.C., Kovács, T., 2018. Surrogate model based optimization of traffic lights cycles and green period ratios using microscopic simulation and fuzzy rule interpolation. Int. J. Artif. Intell. 16 (1), 20–40.

Andonovski, G., Angelov, P., Blažič, S., Škrjanc, I., 2016. A practical implementation of robust evolving cloud-based controller with normalized data space for heat-exchanger plant. Appl. Soft Comput. 48, 29–38.

Andonovski, G., Mušič, G., Blažič, S., Škrjanc, I., 2018. Evolving model identification for process monitoring and prediction of non-linear systems. Eng. Appl. Artif. Intell. 68, 214–221.

Angelov, P., 2002. Evolving Rule-Based Models: A Tool for Design of Flexible Adaptive Systems. Springer-Verlag, Heidelberg, p. 2002.

Angelov, P., 2004. A fuzzy controller with evolving structure. Inf. Sci. 161 (1–2), 21–35.

Angelov, P., Buswell, R.A., Wright, J.A., Loveday, D.L., 2001. Evolving rules-based control. In: Proceedings of EUNITE 2001 Symposium, Tenerife, Spain, pp. 36–41.

Angelov, P., Buswell, R.A., 2002. Identification of evolving fuzzy rule-based models. IEEE Trans. Fuzzy Syst. 10 (5), 667–677.

Angelov, P., Filev, D., 2002. Flexible models with evolving structure. In: Proceedings of 2002 First International IEEE Symposium on Intelligent Systems, Varna, Bulgaria, pp. 28–33.

Angelov, P., Filev, D., 2003. On-line design of Takagi-Sugeno models. In: Bilgiç, T., De Baets, B., Kaynak, O. (Eds.), Fuzzy Sets and Systems - IFSA 2003. In: Lecture Notes in Computer Science, vol. 2715. Springer-Verlag, Berlin, Heidelberg, pp. 576–584.

Angelov, P., Filev, D., 2004. An approach to online identification of Takagi-Sugeno fuzzy models. IEEE Trans. Syst. Man Cybern. B Cybern. 34 (1), 484–498.

Baghbani, F., Akbarzadeh-T, M.-R., Akbarzadeh, A., 2018. Indirect adaptive robust mixed H_2/H_∞ general type-2 fuzzy control of uncertain nonlinear systems. Appl. Soft Comput. 72, 392–418.

Bara, O., Fliess, M., Join, C., Day, J., Djouadi, S.M., 2018. Toward a model-free feedback control synthesis for treating acute inflammation. J. Theor. Biol. 478, 26–37.

Baranyi, P., 2015. TP model transformation as a manipulation tool for qLPV analysis and design. Asian J. Control 17 (2), 497–507.

Baranyi, P., 2016. TP-Model Transformation-Based-Control Design Frameworks. Springer, Cham.

Baranyi, P., Takarics, B., 2014. Aeroelastic wing section control via relaxed tensor product model transformation framework. J. Guid. Control. Dyn. 37 (5), 1671–1678.

Blažič, S., Škrjanc, I., Matko, D., 2014. A robust fuzzy adaptive law for evolving control systems. Evol. Syst. 5 (1), 3–10.

Campi, M.C., Lecchini, A., Savaresi, S.M., 2002. Virtual reference feedback tuning: a direct method for the design of feedback controllers. Automatica 38 (8), 1337–1346.

Campos, V.C.D.S., Torres, L.A.B., Palhares, R.M., 2015. Revisiting the TP model transformation: interpolation and rule reduction. Asian J. Control 17 (2), 392–401.

Castillo, O., Melin, P., 2012. A review on the design and optimization of interval type-2 fuzzy controllers. Appl. Soft Comput. 12 (4), 1267–1278.

Castillo, O., Melin, P., 2014. A review on interval type-2 fuzzy logic applications in intelligent control. Inf. Sci. 279, 615–631.

Castillo, O., Melin, P., Alanis, A., Montiel, O., Sepulveda, R., 2011. Optimization of interval type-2 fuzzy logic controllers using evolutionary algorithms. Soft. Comput. 15 (6), 1145–1160.

David, R.-C., Precup, R.-E., Petriu, E.M., Radac, M.-B., Preitl, S., 2013. Gravitational search algorithm-based design of fuzzy control systems with a reduced parametric sensitivity. Inf. Sci. 247, 154–173.

Filip, F.G., 2008. Decision support and control for large-scale complex systems. Annu. Rev. Control. 32 (1), 61–70.

Fliess, M., 2013. Model-free control. Int. J. Control. 86 (12), 2228–2252.

Formentin, S., Karimi, A., Savaresi, S.M., 2013. Optimal input design for direct data-driven tuning of model-reference controllers. Automatica 49 (6), 1874–1882.

Gao, Z., 2006. Active disturbance rejection control: a paradigm shift in feedback control system design. In: Proceedings of 2006 American Control Conference, Minneapolis, MN, USA, pp. 2399–2405.

Gong, H.-H., Sun, H., Wang, B.-L., Yu, Y., Li, Z., Liao, X.-Z., Liu, X.-D., 2018. Tensor product model-based control for spacecraft with fuel slosh dynamics. Acta Polytech. Hung. 15 (3), 63–80.

Groumpos, P.P., 2018. Intelligence and fuzzy cognitive maps: scientific issues, challenges and opportunities. Studies in Informatics and Control 27 (3), 247–264.

Guerrero, J., Torres, J., Antonio, E., Campos, E., 2018. Autonomous underwater vehicle robust path tracking: generalized super-twisting algorithm and block backstepping controllers. Control Eng. Appl. Inf. 20 (2), 51–63.

Haber, R.E., Alique, J.R., 2004. Nonlinear internal model control using neural networks: an application for machining processes. Neural Comput. Applic. 13 (1), 47–55.

Hedrea, E.-L., Bojan-Dragos, C.-A., Precup, R.-E., Roman, R.-C., Petriu, E.M., Hedrea, C., 2017. Tensor product-based model transformation for position control of magnetic levitation systems. In: Proceedings of 2017 IEEE International Symposium on Industrial Electronics, Edinburgh, UK, pp. 1141–1146.

Hedrea, E.-L., Precup, R.-E., Bojan-Dragos, C.-A., Roman, R.-C., Tanasoiu, O., Marinescu, M., 2018. Cascade control solutions for maglev systems. In: Proceedings of 22nd International Conference on System Theory, Control and Computing, Sinaia, Romania, pp. 20–26.

Herbst, G., 2013. A simulative study on active disturbance rejection control (ADRC) as a control tool for practitioners. Electronics 2, 246–279.

Hjalmarsson, H., Gevers, M., Gunnarsson, S., Lequin, O., 1998. Iterative feedback tuning: theory and applications. IEEE Control. Syst. Mag. 18 (4), 26–41.

Hong, Y.-Y., Nguyen, M.-T., 2018. Optimal design of IT2-FCS-based STATCOM controller applied to power system with wind farms using Taguchi method. IET Gener. Transm. Distrib. 12 (13), 3145–3151.

Kayacan, E., Khanesar, M.A., 2016. Recurrent interval type-2 fuzzy control of 2-DOF heli-copter with finite time training algorithm. IFAC-PapersOnLine 49 (13), 293–299.

Kovács, L., Eigner, G., 2018. Tensor product model transformation based parallel distributed control of tumor growth. Acta Polytech. Hung. 15 (3), 101–123.

Korondi, P., 2006. Tensor product model transformation-based sliding surface design. Acta Polytech. Hung. 3 (4), 23–36.

Leite, D., Palhares, R.M., Campos, V.C.S., Gomide, F.A.C., 2015. Evolving granular fuzzy model-based control of nonlinear dynamic systems. IEEE Trans. Fuzzy Syst. 23 (4), 923–938.

Li, M., Zhu, Y., Hang, K.-M., Yang, L.-H., Hu, C.-X., Mu, H.-H., 2019. Convergence rate oriented iterative feedback tuning with application to an ultraprecision wafer stage. IEEE Trans. Ind. Electron. 66 (3), 1993–2003.

Linda, O., Manic, M., 2011. Uncertainty-robust design of interval type-2 fuzzy logic con-troller for delta parallel robot. IEEE Trans. Ind. Inf. 7 (4), 661–670.

Lughofer, E., Pratama, M., Skrjanc, I., 2018. Incremental rule splitting in generalized evolv-ing fuzzy systems for autonomous drift compensation. IEEE Trans. Fuzzy Syst. 26 (4), 1854–1865.

Mohammadzadeh, A., Kaynak, O., Teshnehlab, M., 2014. Two-mode indirect adaptive control approach for the synchronization of uncertain chaotic systems by the use of a hierarchical interval type-2 fuzzy neural network. IEEE Trans. Fuzzy Syst. 22 (5), 1301–1312.

Nguang, S.-K., Shi, P., 2003. H_∞ fuzzy output feedback control design for nonlinear sys-tems: an LMI approach. IEEE Trans. Fuzzy Syst. 11 (3), 331–340.

Nguyen, A.-T., Taniguchi, T., Eciolaza, L., Campos, V., Palhares, R., Sugeno, M., 2019. Fuzzy control systems: past, present and future. IEEE Comput. Intell. Mag. 14 (1), 1–13.

Oh, S.K., Jang, H.J., Pedrycz, W., 2011. A comparative experimental study of type-1/type-2 fuzzy cascade controller based on genetic algorithms and particle swarm optimization. Expert Syst. Appl. 38 (9), 11217–11229.

Osaba, E., Carballedo, R., Yang, X.-S., Diaz, F., 2016. An evolutionary discrete firefly algo-rithm with novel operators for solving the vehicle routing problem with time windows. In: Yang, X.-S. (Ed.), Nature-Inspired Computation in Engineering. In: Studies in Computational Intelligence, vol. 637. Springer, Cham, pp. 21–41.

Precup, R.-E., Angelov, P., Costa, B.S.J., Sayed-Mouchaweh, M., 2015a. An overview on fault diagnosis and nature-inspired optimal control of industrial process applications. Comput. Ind. 74, 75–94.

Precup, R.-E., David, R.-C., Petriu, E.M., 2017a. Grey wolf optimizer algorithm-based tuning of fuzzy control systems with reduced parametric sensitivity. IEEE Trans. Ind. Electron. 64 (1), 527–534.

Precup, R.-E., David, R.-C., Szedlak-Stinean, A.-I., Petriu, E.M., Dragan, F., 2017b. An easily understandable grey wolf optimizer and its application to fuzzy controller tuning. Algorithms 10 (2), 1–15. https://doi.org/10.3390/a10020068.

Precup, R.-E., David, R.-C., Petriu, E.M., Preitl, S., Radac, M.-B., 2012a. Novel adaptive gravitational search algorithm for fuzzy controlled servo systems. IEEE Trans. Ind. Inf. 8 (4), 791–800.

Precup, R.-E., David, R.-C., Petriu, E.M., Preitl, S., Radac, M.-B., 2012b. Fuzzy control systems with reduced parametric sensitivity based on simulated annealing. IEEE Trans. Ind. Electron. 59 (8), 3049–3061.

Precup, R.-E., David, R.-C., Petriu, E.M., Preitl, S., Radac, M.-B., 2014a. Novel adaptive charged system search algorithm for optimal tuning of fuzzy controllers. Expert Syst. Appl. 41 (4), 1168–1175. part 1.

Precup, R.-E., David, R.-C., Petriu, E.M., Preitl, S., Radac, M.-B., 2013a. Fuzzy logic-based adaptive gravitational search algorithm for optimal tuning of fuzzy controlled servo systems. IET Control Theory Appl. 7 (1), 99–107.

Precup, R.-E., David, R.-C., Petriu, E.M., Radac, M.-B., Preitl, S., Fodor, J., 2013b. Evolutionary optimization-based tuning of low-cost fuzzy controllers for servo systems. Knowl.-Based Syst. 38, 74–84.

Precup, R.-E., Filip, H.-I., Radac, M.-B., Petriu, E.M., Preitl, S., Dragos, C.-A., 2014b. Online identification of evolving Takagi-Sugeno-Kang fuzzy models for crane systems. Appl. Soft Comput. 24, 1155–1163.

Precup, R.-E., Hellendoorn, H., 2011. A survey on industrial applications of fuzzy control. Comput. Ind. 62 (3), 213–226.

Precup, R.-E., Petriu, E.M., Radac, M.-B., Preitl, S., Fedorovici, L.-O., Dragos, C.-A., 2015b. Cascade control system-based cost effective combination of tensor product model transformation and fuzzy control. Asian J. Control 17 (2), 381–391.

Precup, R.-E., Preitl, S., Rudas, I.J., Tomescu, M.L., Tar, J.K., 2008a. Design and experiments for a class of fuzzy controlled servo systems. IEEE/ASME Trans. Mechatron. 13 (1), 22–35.

Precup, R.-E., Preitl, S., Tar, J.K., Tomescu, M.L., Takács, M., Korondi, P., Baranyi, P., 2008b. Fuzzy control system performance enhancement by iterative learning control. IEEE Trans. Ind. Electron. 55 (9), 3461–3475.

Precup, R.-E., Radac, M.-B., Roman, R.-C., Petriu, E.M., 2017c. Model-free sliding mode control of nonlinear systems: algorithms and experiments. Inf. Sci. 381, 176–192.

Precup, R.-E., Sabau, M.-C., Petriu, E.M., 2015c. Nature-inspired optimal tuning of input membership functions of Takagi-Sugeno-Kang fuzzy models for anti-lock braking systems. Appl. Soft Comput. 27, 575–589.

Precup, R.-E., Teban, T.-A., Petriu, E.M., Albu, A., Mituletu, I.-C., 2018. Structure and evolving fuzzy models for prosthetic hand myoelectric-based control systems. In: Proceedings of 26th Mediterranean Conference on Control and Automation, Zadar, Croatia, pp. 625–630.

Precup, R.-E., Tomescu, M.L., Preitl, S., Petriu, E.M., Fodor, J., Pozna, C., 2013c. Stability analysis and design of a class of MIMO fuzzy control systems. J. Intell. Fuzzy Syst. 25 (1), 145–155.

Precup, R.-E., Voisan, E.-I., Petriu, E.M., Radac, M.-B., Fedorovici, L.-O., 2016. Gravitational search algorithm-based evolving fuzzy models of a nonlinear process. In: Filipe, J., Madani, K., Gusikhin, O., Sasiadek, J. (Eds.), Informatics in Control, Automation and Robotics. In: Lecture Notes in Electrical Engineering, vol. 383. Springer, Cham, pp. 51–62.

Preitl, S., Precup, R.-E., Fodor, J., Bede, B., 2006. Iterative feedback tuning in fuzzy control systems. Theory and applications. Acta Polytech. Hung. 3 (3), 81–96.

Purcaru, C., Precup, R.-E., Iercan, D., Fedorovici, L.-O., David, R.-C., Dragan, F., 2013. Optimal robot path planning using gravitational search algorithm. Int. J. Artif. Intell. 10 (S13), 1–20.

Radiša, R., Dučić, N., Manasijević, S., Marković, N., Ćojbašić, Ž., 2017. Casting improvement based on metaheuristic optimization and numerical simulation. Facta Universitatis, Series Mechanical Engineering 15 (3), 397–411.

Roman, R.-C., Precup, R.-E., David, R.-C., 2018. Second order intelligent proportional-integral fuzzy control of twin rotor aerodynamic systems. Procedia Comput. Sci. 139, 372–380.

Roman, R.-C., Precup, R.-E., Radac, M.-B., 2017. Model-free fuzzy control of twin rotor aerodynamic systems. In: Proceedings of 2017 25th Mediterranean Conference on Control and Automation, Valletta, Malta, pp. 559–564.

Rong, H.-J., Yang, Z.-X., Wong, P.K., Vong, C.M., Zhao, G.-S., 2017. Self-evolving fuzzy model-based controller with online structure and parameter learning for hypersonic vehicle. Aerosp. Sci. Technol. 64, 1–15.

Sadeghi-Tehran, P., Cara, A.B., Angelov, P., Pomares, H., Rojas, I., Prieto, A., 2012. Self-evolving parameter-free rule-based controller. In: Proceedings of 2012 World Congress on Computational Intelligence, Brisbane, Australia. pp. 754–761.

Safaei, A., Mahyuddin, M.N., 2018. Optimal model-free control for a generic MIMO non-linear system with application to autonomous mobile robots. Int. J. Adapt Control Signal Process. 32 (6), 792–815.

Sirbu, I., Dumitrache, I., 2017. A conceptual framework for artificial creativity in visual arts. Int. J. Comput. Commun. Control 12 (3), 381–392.

Son, D.-G., Choi, H.-C., 2018. Iterative feedback tuning of the proportional-integral-differential control of flow over a circular cylinder. IEEE Trans. Control Syst. Technol. 1–12. https://doi.org/10.1109/TCST.2018.2828381.

Szöllősi, A., Baranyi, P., 2018. Influence of the tensor product model representation of qLPV models on the feasibility of linear matrix inequality based stability analysis. Asian J. Control 20 (1), 531–547.

Vaščák, J., Rutrich, M., 2008. Path planning in dynamic environment using fuzzy cognitive maps. In: Proceedings of 6th International Symposium on Applied Machine Intelligence and Informatics, Herľany, Slovakia, pp. 5–9.

Vrkalović, S., Lunca, E.-C., Borlea, I.-D., 2018. Model-free sliding mode and fuzzy controllers for reverse osmosis desalination plants. Int. J. Artif. Intell. 16 (2), 208–222.

Wang, T.-C., Tong, S.-C., Yi, J.-Q., Li, H.-Y., 2015. Adaptive inverse control of cable-driven parallel system based on type-2 fuzzy logic systems. IEEE Trans. Fuzzy Syst. 23 (5), 1803–1816.

Yu, X., Hou, Z., Zhang, X., 2018. Model-free adaptive control for a vapour-compression refrigeration benchmark process. IFAC-PapersOnLine 51 (4), 527–532.

Za'in, C., Pratama, M., Lughofer, E., Anavatti, S.G., 2017. Evolving type-2 web news mining. Appl. Soft Comput. 54, 200–220.

Zdešar, A., Dovžan, D., Škrjanc, I., 2014. Self-tuning of 2 DOF control based on evolving fuzzy model. Appl. Soft Comput. 19, 403–418.

Zeng, X.-J., Singh, M.-G., 1996. Approximation accuracy analysis of fuzzy systems as function approximators. IEEE Trans. Fuzzy Syst. 4 (1), 44–63.

Zhu, Y., Hou, Z., Qian, F., Du, W., 2017. Dual RBFNNs-based model-free adaptive control with Aspen HYSYS simulation. IEEE Trans. Neural Netw. Learn. Syst. 28 (3), 759–765.

INDEX

Note: Page numbers followed by *f* indicate figures and *t* indicate tables.

A

Adaptive charged system search
  algorithms, 86–87
  accuracy rate, 119*t*
  convergence speed, 117*t*
  exploitation ability, 88
  flowchart, 90*f*
  objective function evolution *vs.* iteration
    index, 92, 92*f*
  stages, 89–90
  Takagi-Sugeno PI-FC tuning parameters,
    92, 92*f*
  vector solution, 92, 93*f*
Adaptive gravitational search algorithms,
  81–82
  accuracy rate, 119*t*
  convergence speed, 117*t*
  elaboration, 83
  engagement, 82
  evaluation, 84
  explanation, 83
  exploration, 82
  flowchart, 84, 85*f*
  fuzzy logic-based, 92–99
  objective function evolution *vs.* iteration
    index, 84–85, 86*f*
  Takagi-Sugeno PI-FC tuning parameters,
    85, 86*f*
  vector solution, 85–86, 87*f*
Ant colony optimization (ACO), 93–94
Antecedent, 3, 17

C

Center of gravity (CoG) method, 24, 27
Charged particles (CPs), 67, 86–87
Charged system search (CSS) algorithms, 67.
  See also Adaptive charged system
    search algorithms
  accuracy rate, 119*t*
  convergence speed, 117*t*
  flowchart, 70, 71*f*

objective function evolution *vs.* iteration
  index, 70–71, 72*f*
optimization problems, 69–70
for Takagi-Sugeno PI-FCs, 70
tuning parameters, 71, 72*f*
vector solution, 71, 73*f*
Concentration operator (CON*(A)*), 12
Consequent, 3, 17
Control error, 4, 13, 41–42
Controlled process (CP), 1–2, 120
Convergence analysis theorem, 95
Convergence speed, 116–117

D

Defuzzification, 5
  center of gravity (CoG) method, 25
  mean of maxima (MoM) method, 24–25
  parameters, 26–28
  in Takagi-Sugeno fuzzy models, 29
  weighted average method, 25–26
Descriptors, 9
Dilation operator (DIL*(A)*), 12
Discrete-time objective functions, 42–43

E

Evolving fuzzy systems, 127–131
Extended symmetrical optimum (ESO)
  method, 45–47, 121
Extremity problem, 26

F

First-order Pade approximation, 35
First-order Sugeno fuzzy model, 28–29,
  30*f*
5E learning cycle, 82
Fuzzification, 5, 13–17, 14*f*
Fuzzy congruence, 9
Fuzzy control/fuzzy logic control, 2
  applications, 2–3, 6–7
  auxiliary variables, 4
  evolution, 3–4
  features, 3

Fuzzy control/fuzzy logic control
(*Continued*)
principle, 4, 4*f*
rule base, 3
Fuzzy controllers (FC), 2–3
advantages, 6
with dynamics, 33–41
optimal tuning, 5, 133–134
without dynamics, 31–33
Fuzzy control rules, 5
Fuzzy control signal, 5
Fuzzy control systems (FCSs), 3
evolution, 127–131
performance, 45, 46*f*
real-time experiment, 119–121
structure, 41–42, 41*f*
Fuzzy information, 7–8
Fuzzy logic-based adaptive GSAs, 92–99
accuracy rate, 119*t*
convergence speed, 118*t*
Fuzzy propositions, 17
Fuzzy sets
congruence and strictly fuzzy
congruence, 9
descriptors, 9
membership function, 7–8
modifiers, 11–13
operators, 9–11
Fuzzy set (FS) theory, 2
Fuzzy subset, 9–11

G

Gaussian membership function, 8
Generalized bell-shaped membership
function, 8
Gravitational search algorithm (GSA), 62.
See also Adaptive gravitational search
algorithms
accuracy rate, 119*t*
convergence speed, 117*t*
flowchart, 65–66, 65*f*
gravitational and inertial masses, 64
objective function evolution *vs.* iteration
index, 66, 67*f*
state-space equations, 64
stochastic characteristic, 64
for Takagi-Sugeno PI-FCs, 63

tuning parameters, 66, 67*f*
vector solution, 66, 68*f*
Gray wolf optimizer (GWO) algorithm
accuracy rate, 119*t*
applications, 72–73
convergence speed, 117*t*
exploitation stage, 74
flowchart, 75–76, 76*f*
objective function evolution *vs.* iteration
index, 77–78, 77*f*
operating mechanism, 73
Takagi-Sugeno PI-FC tuning parameters,
77–78, 77*f*
vector solution, 75, 77–78, 78*f*

H

High-order singular value decomposition
(HOSVD), 125–127
Hybrid gray wolf optimization-particle
swarm optimization algorithms, 108
accuracy rate, 119*t*
convergence speed, 118*t*
flowchart, 110–111, 110*f*
objective function evolution *vs.* iteration
index, 111, 112*f*
search process, 109
Takagi-Sugeno PI-FC tuning
parameters, 111, 112*f*
vector solution, 111, 113*f*
Hybrid particle swarm optimization-
gravitational search algorithms
accuracy rate, 119*t*
applications, 103–104
convergence speed, 118*t*
exploitation and exploration, 104–105
flowchart, 105–106, 105*f*
objective function evolution *vs.* iteration
index, 106, 107*f*
operating mechanism, 104
Takagi-Sugeno PI-FC tuning parameters,
106, 107*f*
vector solution, 108, 112*f*

I

IF-THEN rules, 3, 5, 17
Incremental online identification
algorithm, 127, 129–131

Indirect learning mechanism, 130–131
Inference mechanism, 5
    MAX-MIN, 20–21, 22*f*
    MAX-PROD, 19–20
    SUM-PROD, 19–20
    in Takagi-Sugeno fuzzy models, 22–23,
        29
Inference module, 17–24
Inference table, 17, 21, 21*t*
Input normalization, 15
Intelligent control systems, 3–4
Intensification operator (INT*(A)*), 12

L

Linear time-invariant (LTI) system, 94, 96
Linguistic terms (LTs), 13–14
Linguistic variable (LV), 13–14

M

Mamdani fuzzy controllers, 4–6
Mamdani fuzzy rule bases, 18
Mamdani PI-fuzzy controllers
    design, 39
    membership functions, 36, 36*f*
Mamdani's fuzzy complement operator, 10
MAX-MIN inference mechanism, 20–21,
        22*f*
MAX operator, 10
MAX-PROD inference mechanism, 19–20
Mean of maxima (MoM) method, 24–25, 27
Membership functions, 7–8
    of Mamdani PI-fuzzy controllers, 36, 36*f*
    of Takagi-Sugeno PI-FCs, 40, 40*f*
Metaheuristic algorithms/metaheuristics,
        6, 108
MIN operator, 10, 17
MISO nonlinear proportional FCs, 32
Modal equivalence principle, 38, 41
Modal value, 8
Modification operators/modifiers, 11–13
Multiple input-multiple output (MIMO),
        3, 94

N

n-ary fuzzy relation, 12
Nonlinear (NL) feedback block, 95

O

Objective functions, 42–43
    adaptive charged system search
        algorithms, 91–92
    charged system search (CSS) algorithms,
        70–71, 72*f*
    convergence, 43, 49
    fuzzy logic-based adaptive GSAs, 98–99
    gravitational search algorithm, 66
    gray wolf optimizer (GWO) algorithm,
        74, 77–78, 77*f*
    GWOPSO algorithm, 111, 111*t*
    PSO algorithm, 60, 61*f*
    variables, 48
Output denormalization, 15

P

Particles, 63
Particle swarm optimization (PSO), 55–56.
        See also Hybrid particle swarm
        optimization-gravitational search
        algorithms
    accuracy rate, 119*t*
    binary version, 59
    characteristic, 59
    computational attributes, 56
    convergence speed, 117*t*
    flowchart, 59
    objective function evolution *vs.* iteration
        index, 60, 61*f*
    principles, 56
    star-type topology, 57–58
    for Takagi-Sugeno PI-FCs, 57
    tuning parameters, 60, 61*f*
    vector solution, 60, 62*f*
Performance index. *See* Convergence speed
Popov's hyperstability analysis, 93–96
PROD operator, 10, 19–20
Proportional-derivative fuzzy controllers
        (PD-FCs), 121–122
Proportional-integral- derivative (PID)
        controllers, 35–36
Proportional-integral-derivative fuzzy
        controllers (PID-FCs)
    design approach, 122
    structure, 121–122, 122*f*
    uses, 121

Proportional-integral fuzzy controllers (PI-FCs), 35–37, 36f, 40. *See also* Takagi-Sugeno PI-FCs
PSO. *See* Particle swarm optimization (PSO)
Pulse width modulation (PWM) signals, 47–48

Q

Quasi-linear parameter-varying (qLPV), 126–127

R

Rule base, 5
    decision table, 21
    definition, 23–24
    of SITO fuzzy block, 96–97
    of Takagi-Sugeno fuzzy models, 128

S

Scaling, 15, 26
Scheduling variables, 4
Set-point filter FCS, 41–42, 41f
Single input-two output (SITO) fuzzy block, 96, 97f
Singleton, 7–8
Singleton membership function, 8
SISO nonlinear proportional FC, 31–33
Standard deviation, 118
State-space equations, 64
Strictly fuzzy congruence, 9
Sugeno models. *See* Takagi-Sugeno fuzzy model
SUM operator, 10, 98
SUM-PROD inference mechanism, 19–20
Swarm intelligence, 56

T

Takagi-Sugeno fuzzy controllers, 7, 22
Takagi-Sugeno fuzzy model, 28–31, 124–125
    operating principle, 29, 30f
    rule base, 128
Takagi-Sugeno fuzzy rules, 22
Takagi-Sugeno PI-FCs

accuracy rate, 118–119, 119t
adaptive charged system search algorithms, 92, 92f
adaptive gravitational search algorithms, 85, 86f
charged system search (CSS) algorithms, 70
convergence speed, 116–118
decision table, 40, 40t
design approach, 39, 46–47
experimental setup, 47, 48f
gravitational search algorithm, 63
gray wolf optimizer (GWO) algorithm, 77–78, 77f
GWOPSO algorithm, 111
input membership functions, 40, 40f
objective functions, average values, 116
PSO algorithm, 57
PSOGSA algorithm, 106
Tensor product (TP)-based model transformation
    advantage, 125
    definition, 124
    functions, 124–125
    uses, 125–127
Trapezoidal membership function, 8
Triangular membership function, 8
Triangular norms (t-norms), 10
Tsukamoto fuzzy models, 30, 31f
Tustin's method, 36–37, 46, 122–123
Two degrees-of-freedom (2-DOF), 41–42, 45, 123
Type-2 fuzzy logic systems, 93–94, 132
    design and tuning, 123–124
    operation principle, 123
    structure, 123, 124f

W

Weighted average method, 25–26, 129–130
Weighted average operator, 29–30

Z

Zero-order Sugeno fuzzy model, 28–29, 31

Printed in the United States
By Bookmasters